Innovative Renewable Energy

Series editor
Ali Sayigh
World Renewable Energy Congress, Brighton, UK

The primary objective of this book series is to highlight the best-implemented worldwide policies, projects and research dealing with renewable energy and the environment. The books will be developed in partnership with the World Renewable Energy Network (WREN). WREN is one of the most effective organizations in supporting and enhancing the utilisation and implementation of renewable energy sources that are both environmentally safe and economically sustainable. Contributors to books in this series will come from a worldwide network of agencies, laboratories, institutions, companies and individuals, all working together towards an international diffusion of renewable energy technologies and applications. With contributions from most countries in the world, books in this series will promote the communication and technical education of scientists, engineers, technicians and managers in this field and address the energy needs of both developing and developed countries.

Each book in the series will contain contributions from WREN members and will cover the most-up-to-date research developments, government policies, business models, best practices, and innovations from countries all over the globe. Additionally, the series will publish a collection of best papers presented during the annual and bi-annual World Renewable Energy Congress and Forum each year.

More information about this series at http://www.springer.com/series/15925

Ali Sayigh
Editor

Seaside Building Design: Principles and Practice

Buildings in Maritime Zones

Editor
Ali Sayigh
World Renewable Energy Congress and Network
Brighton, UK

ISSN 2522-8927 ISSN 2522-8935 (electronic)
Innovative Renewable Energy
ISBN 978-3-319-88517-9 ISBN 978-3-319-67949-5 (eBook)
https://doi.org/10.1007/978-3-319-67949-5

Printed on acid-free paper

This Springer imprint is published by Springer Nature
The registered company is Springer International Publishing AG
The registered company address is: Gewerbestrasse 11, 6330 Cham, Switzerland

Brighton Pavilion – UK

Preface

This book, *Seaside Building Design: Principles and Practice*, examines the major features of architecture and building design in various coastal zones, primarily in moderate climates with high or average humidity. Demand for seaside living is often at a premium. Luxury villas and resorts, expensive residential developments, high-rise office buildings, and apartment blocks proliferate in harbors, seaports, and coastal regions. Meeting this demand is challenging, particularly in terms of sustainable development. This is a result of climate change, which has engendered extreme weather patterns that result in high winds, heat waves, storms, floods, and landslides. So far, this has not significantly deterred people from living by the sea.

It is essential that architects, developers, planners, and builders integrate these new environmental problems into their designs and development plans, while continuing to capitalize on the many benefits of living in coastal zones. This will mean greater care being taken regarding site location and building materials, use of natural ventilation, shading devices, and appropriate levels of maintenance.

Brighton, UK
2017

Ali Sayigh

Contents

Chapter 1
Introduction

Ali Sayigh

Generally, it is found that the coastal climate is less severe than that of the inland and is therefore considered preferable for living. Every meteorological zone is unique which must be taken into consideration by architects.

"Living by the seaside boosts mental health, makes people happier and more relaxed," according to new research, by Sarah O'Grady, published in the *Sunday Express* on 29 April 2016." O'Grady points out that American analysis of New Zealand data found residents whose properties had an ocean view were happier than their landlocked neighbors and is the first report to find a link between health and the visibility of water. The research showed how the sound of waves affected the wave patterns in the brain lulling a person into a relaxed state which can help rejuvenate the mind and body. Professor Amber Pearson, of Michigan State University in the USA, said: "Increased views of blue space is significantly associated with lower levels of psychological distress. However, we did not find that with green space."

Bearing this in mind, nine top architects from various global locations were commissioned to write their design experience of seaside buildings. This is a novel approach aimed at all architects and builders who are designing and working in projects for the seaside environment.

Chapter 2: *The Beach House at Bexhill, England*, by Architect Nazar Sayigh

This is a unique house design incorporating the old existing dwelling within the structure of the new design. Sayigh strived to devise a property that would fit its environs, reflecting the cool colors of the sea and shingle coast line, while hinting retrospectively, in exterior appearance, at its sustainable construction. It is an attempt to blend modernist Californian beach design within the constraints required by the local Sussex seaside environment to give a comfortable domestic space reflecting the natural pebble beach coast in which it is situated.

A. Sayigh (✉)
World Renewable Energy Congress and Network, Brighton, UK
e-mail: asayigh@wrenuk.co.uk

A. Sayigh (ed.), *Seaside Building Design: Principles and Practice*,
Innovative Renewable Energy, https://doi.org/10.1007/978-3-319-67949-5_1

Chapter 3: *Natural Ventilation in Hot Seaside Urban Environments*, by Prof. Khaled A. Al-Sallal and Mrs. Amira R. AbouElhamd, UAE

The article considers the growing population migration from rural to urban regions around the world; for example, in 1950 the urban population was 30% of the global population while in 2014 this had risen 54% (3.9 billion). This growth was associated with increased demand for energy and led to excessive greenhouse gases (GHG) emissions and pollution. In the UAE, the building sector consumes more than 48% of the total used energy. Obviously, the bulk of this energy is used for air conditioning for cooling. Therefore, using natural ventilation would not only be healthier in most cases but also reduce the use of fossil fuels. This chapter uses the city of Dubai as its prime example.

Chapter 4: *Architecture and the Sea: The Situation in the Netherlands*, by Prof. Dr. Wim Zeiler from the Netherlands

In this article, Prof. Zeiler emphasizes the vulnerability of the Netherlands as a result of having 26% of the land mass below sea level while 60% is under threat of flood; indeed, since the fourteenth century, windmills have been pumping water off thc land to keep it dry. Half of the Dutch people, 4 million, lives below sea level, so architects are very innovative and obsessed by the problems associated with water and growing concerns with the climate. Sustainability determines the ecological value and its relationship with the environment and involves using the 3P approach: people, planet, and profitability. Several examples of buildings at the seaside are discussed and conclusions are drawn.

Chapter 5: *Sustainability Measures of Public Buildings in Seaside Cities: The New Library of Alexandria (New Bibliotheca Alexandrina), Egypt,* by Prof. Dr. Mohsen M. Aboulnaga, Cairo University, Egypt

The article describes in details how this iconic building was designed with all aspects of sustainability, comfort, and functionality. It follows the norms of effective and efficient framework for public library design.

To design buildings at a seaside location, the effect of climate change and sustainability have to be considered. Three libraries are compared: one in Boston, USA (John F. Kennedy); second in Copenhagen, Denmark (The Royal Danish Library); and the third in Melbourne, Australia (The Docklands Library). Each library's feature was fully compared with reference to concept and form, spatial experience, building conditions, and sustainability. Daylight factor was one of the crucial elements in the overall design.

Chapter 6: *Seaside Buildings in Portugal*, by Prof. Manuel Correia Guedes

The author outlines the importance of the sea to Portugal since it is surrounded by the sea on two sides. It is estimated that more than 80% of the Portuguese population lives in coastal areas.

Portugal enjoys reasonable level of tourism throughout the year.

In the case study, several locations are discussed in terms of building design suitability and its environmental impact. Renewable energy utilization is another area discussed in depth as well as various recommendations for the mitigation of the effects of wind and sea current erosion along the coast and in settlements. A concise description of and means for achieving urban sustainable development with various examples in Portugal is presented. Espinho was chosen as a case study, and full analysis was made into how architects planned, designed, and constructed it.

Chapter 7: *Climate Adaptive Design on the Norwegian Coast*, by Prof. Luca Finocchiaro, Norway

Formerly Norwegian coastal construction was limited to seasonal fishing villages in a variety of coastal climates which gave rise to different design solutions.

With the transition of the fishing industry from a family-driven activity to the domain of large companies, fishing villages have transitioned into an era when most of their building stock has been converted into housing or devoted to the tourism industry. The author shows how site-specific solutions have been developed in order to adapt to the various climate zones of the Norwegian coast, especially bearing in mind climate change as a result of global warming.

Chapter 8: *Green Design for a Smart Island: Green Infrastructure and Architectural Solutions for Ecotourism in Mediterranean Areas*, by Prof. Antonella Trombadore, Italy

The author defines a green building as being one which has higher efficiency, with less energy, water, and material usage. It has a minimal health effect on the inhabitants and the environment by having improved design and construction and operation as well as maintenance plus the removal of any excessive materials throughout the building life cycle. The author's approach involved discussing five elements: architecture and climate condition, inclusiveness and change, identity and competitiveness, urban transformation and environmental quality, and innovation and tradition. Ecotourism was addressed through the extensive discussion of six pilot projects.

Chapter 9: *Twenty-Four Bioclimatic Dwellings for the Island of Tenerife: Twenty Years Later*, by Dr. Judit Lopez-Besora and Prof. Helena Coch Roura

The article was a result of a competition for ideal single home by the sea organized by the International Union of Architects (UIA) 1996. There were 397 submissions from 38 countries. Full details of the competition and the results two decades later are documented by the authors. Enough collected data was collected to carry out a full analysis to assess the implemented solutions. Case studies, measurements, and literature reviews during these years have provided a useful perspective for the assessment of the first attempts to implement the concept of a bioclimatic project in Tenerife. In the conclusion, the authors discuss the evolution of vernacular architecture and what has been learnt from this study. The most successful design options implemented in the houses are related with the presence of wind protection, courtyards, and facade filters to regulate the connection with the exterior. These approaches are very similar to those of traditional architecture.

Chapter 10: *Design of Seaside Buildings in China*, by Prof. Marco Sala and Prof. Antonella Trombadore, Italy

This a study of the Green Eco Solar Buildings and Sea Side Village Project, developed within the ABITA Research Centre of the University of Florence, for a seaside area in China. It was developed together with the Dafeng Lida Group from Beijing with the aim being to harmonize a traditional Chinese approach with that of sustainable technologies and renewable energy integration in building. In order to achieve green solar buildings, the following five elements were used:

1. Ecological optimization of Green Eco Solar Building project according to the urban development and landscape planning
2. Concept design of Green Eco Solar buildings (homes, hotels, and infrastructure)
3. Building integration of highly efficient strategies, high-tech solutions, and construction materials deriving from recycling and reuse program
4. Sustainable approach to foster local economic development
5. Integrated activities to optimize the use of renewable energy

The second project was to design green villas and solar buildings. In this task, the following strategies were adopted:

Energy management – smart grid and architectural integration of renewable energy
The use of local construction materials and high technology
Water saving and optimization

Chapter 2
The Beach House at Bexhill, England, UK

Nazar Sayigh

2.1 Setting the Scene

Inspiration (Fig. 2.1) for the beach house stemmed primarily from a love of modernist design. Though the client needed the property to fulfil their requirements in terms of internal flexibility, their appreciation of coastal builds from earlier in the twentieth century meant we had more freedom to make daring design decisions without compromising on the their practical demands. Particularly significant in our examination of modernist American beach and holiday properties was the work of Pierre Koenig, Paul Rudolph and Craig Ellwood, all of whom sought to create seamlessly flowing designs whilst working with their sites' locality to form builds which, despite their radical modernity, did not look incongruous in their often rural or semirural Californian surroundings; see Fig. 2.2.

An important example, which uses many of the same principles which we sought to put into practice in Herbrand Walk, is Rudolph's Milam Residence. Rudolph was already known for his ingenious methods of construction – in the so-called Cocoon House, he employed a spray on process used originally by the US navy in the disposal of disused ships to create a complex concave roof system. Following on, the Milam property provided the perfect balance between shielding its inhabitants from the elements and celebrating the location's natural virtues. Constructed using hollow concrete blocks, which form the frames for large floor to ceiling windows, the materials ensured that the interior of the house was provided with shelter and shade from the Florida heat, with the windows not only supplying extensive views of the coast but also monopolizing on the sea breeze as a natural method for cooling the property's inner rooms. The result is an elegant seaside facade of stacked rectangles with a simultaneously functional interior (Fig. 2.3).

N. Sayigh (✉)
Glas Architects, Mezzanine Offices, John Trundle Court, Barbican, London EC2Y 8DJ, UK
e-mail: nazar@glasarchitects.co.uk

A. Sayigh (ed.), *Seaside Building Design: Principles and Practice*,
Innovative Renewable Energy, https://doi.org/10.1007/978-3-319-67949-5_2

Fig. 2.1 Beach house at Bexhill

Fig. 2.2 Typical example of a modernist beach house located in California – Pierre Koenig

Fig. 2.3 The 'Cocoon House', 1951 – Paul Rudolph

Similarly, Craig Ellwood's Case Study Houses explored ideas of environmental and residential continuity. In his Case Study House 18, or 'Fields House', Beverley Hills, he retained a stunningly simple exterior design. The intention was to maintain focus on flexibility of usage – the property's almost entirely open plan interior ensured that none of the spaces within the house were limited to a single purpose allowing maxim usability whilst capturing the subtle beauty of modernist principles. His philosophy is helpful here: 'The essence of architecture is the interrelation and interaction of mass, space, plane and line. The purpose of architecture is to enrich the joy and drama of living. The spirit of architecture is its truthfulness to itself: its clarity and logic with respect to its materials and structure'.

This triplicate of essence, purpose and spirit is an eloquent representation of our own focus in creating the property at Herbrand Walk, where we sought to render the space, materials, location and usage inseparable and harmonious; see Figs. 2.4, 2.5 and 2.6.

No beach house in comparison with Californian mid-century modernism would be complete without mention of the properties erected at Sea Ranch, just north of San Francisco. Here, the architects sought to create harmonious structures that complimented and enhanced their idyllic seafront location.

Fig. 2.4 The Milam Residence, 1961 – Paul Rudolph

Fig. 2.5 Case Study House 18 or 'Fields House', 1958 – Craig Ellwood

Fig. 2.6 Sea Ranch, San Francisco – California

Of particular relevance to our project was their careful use of timber both externally and throughout the interiors. The choice of simple timber plank weather boards, left untreated to oxidize over time, provided a compelling visual language referencing the sea worn and weathered driftwood commonly found washed up on the sandy beaches below.

Sea Ranch's embrace of early sustainable principles was also to be of inspiration; the use of cross-ventilation in lieu of mechanical air conditioning, tough self-protecting natural materials, solar shading, deep overhanging south-facing eaves and carful plot orientation all contributed to the success of these iconic seaside dwellings.

2.2 History of the Site

The site itself, located at 16 Herbrand Walk, Cooden Beach, Bexhill on Sea, was purchased in 2000 by our clients (property professionals with a growing young family). Standing at 0.22 acres, at this early stage, the site also included a 1970s house which had fallen into disrepair, comprising of three storeys and some additional outdoor space (Fig. 2.7).

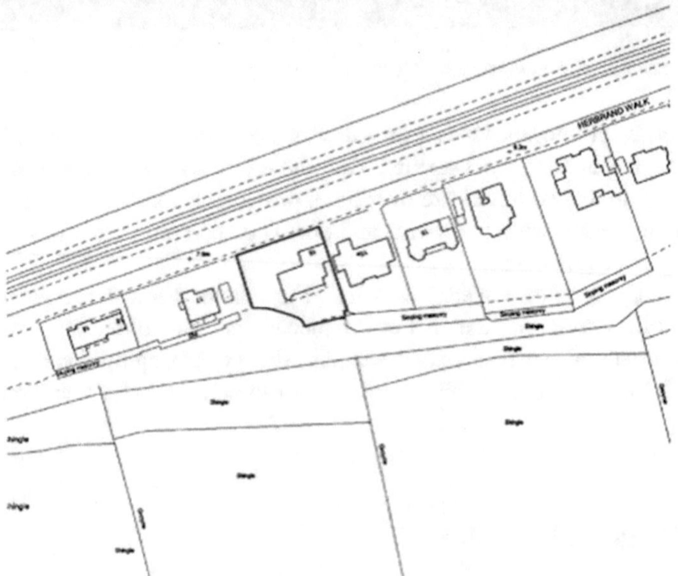

Fig. 2.7 Location of house [edged in red] on Cooden Beach – Bexhill (Aerial view courtesy Apple Maps)

Fig. 2.8 View of the original house

This original property formed part of a row of already established houses built up over time, with designs ranging from the early twentieth century to the 1960s – all of the homes in the row were provided with direct access to the beach. Our initial aim had been to demolish the original house in its entirety and build a new property from scratch. See Figs. 2.8 and 2.9.

On further consideration, we decided instead to strip back the existing home and absorb it fully within the structure of the new design, creating a build, which successfully combined old and new whilst retaining the desired aesthetic. Not only did this ensure a local continuity, but it was also a better option environmentally, allowing the scheme to make full use of the materials already available, as well as contributing added elements to create the final, cohesive structure as in Fig. 2.10.

The location of the site presented some unique difficulties. Unlike the California homes which provided our inspiration, the conditions surrounding our own build were likely to be temperamental; the property would not enjoy the steady stillness and sunshine of its American counterparts. So we chose to wrap both the old and new buildings in an external timber cladding system, with deep-set balconies, providing the double purpose of bringing the finished property together into one house and protecting against the harsh elements of the British coastline and its potentially blustery, wet conditions. See Fig. 2.11.

The spectacular views out to sea from the south and south-west facades could be enjoyed fully without concern regarding the natural environs. So the design sought to maximize the beauty and benefits of the location whilst minimizing its inconveniences (Figs. 2.12, 2.13 and 2.14).

Fig. 2.9 Another view of the original house

Fig. 2.10 Combining the old and the new design to achieve a better result

Fig. 2.11 The use of an external timber cladding system

Fig. 2.12 View of south-west corner of house – illustrating the deep eaves, solar shading and materials

Fig. 2.13 View of the top floor deck due west towards Beachy Head in the far distance

Fig. 2.14 Exterior view of south elevation showing deep eaves, cantilevered corner ceiling to ground floor living room and external pool and terracing

Fig. 2.15 View of house due East as seen from the rear access road – Herbrand Walk (Image courtesy Google Street View)

The placement of the access road so close to the rear of the house, and the main coastal rail line behind that, meant that we were also forced to create a purposefully robust entrance elevation, presenting a significantly more 'private' outlook to this aspect of the house, in contrast with the open south-facing elevation. See Figs. 2.15 and 2.16.

2.3 Design Decisions

Many of the decisions we made in terms of the design, particularly with regard to the interior, revolved around creating a useable, comfortable but also remarkable domestic space, which would suit the multifaceted needs of family life; see Fig. 2.17.

To allow the beauty of the outside environment to seep into the inside of the house, we decided to include full-height south-facing windows on all floors; this not only allows the inhabitants to make the most of the views at all levels but also encourages light to flood through the property – again an environmentally conscious decision which meant that electric lighting would need to be used less frequently during daylight hours. To counter any unpleasant direct sunlight, we also included overhanging eaves which produce areas of shade from the summer glare as well as a protective buffer against wintertime elements.

The top floor of the house has dual functions as both a recreational area and a viewing platform, making the most of the panoramic prospects its windows provide. To enhance this further, we installed a fully folding glazing system which opens the space out onto the terrace area, resulting in a flowing relationship between interior and exterior. Likewise, the ground floor glazing folds back completely at one corner

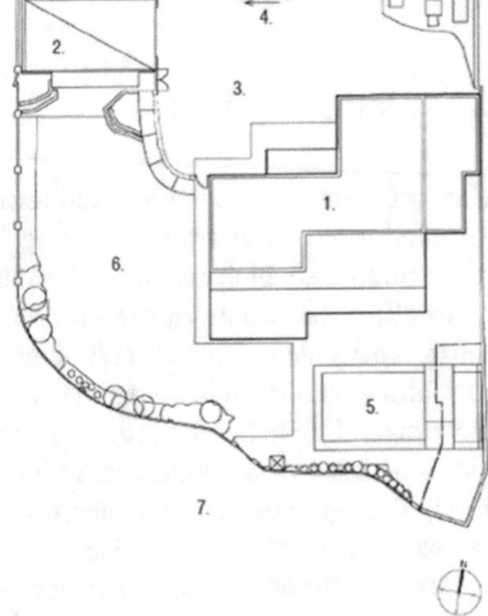

Fig. 2.16 North facing elevation of house highlighting main entrance

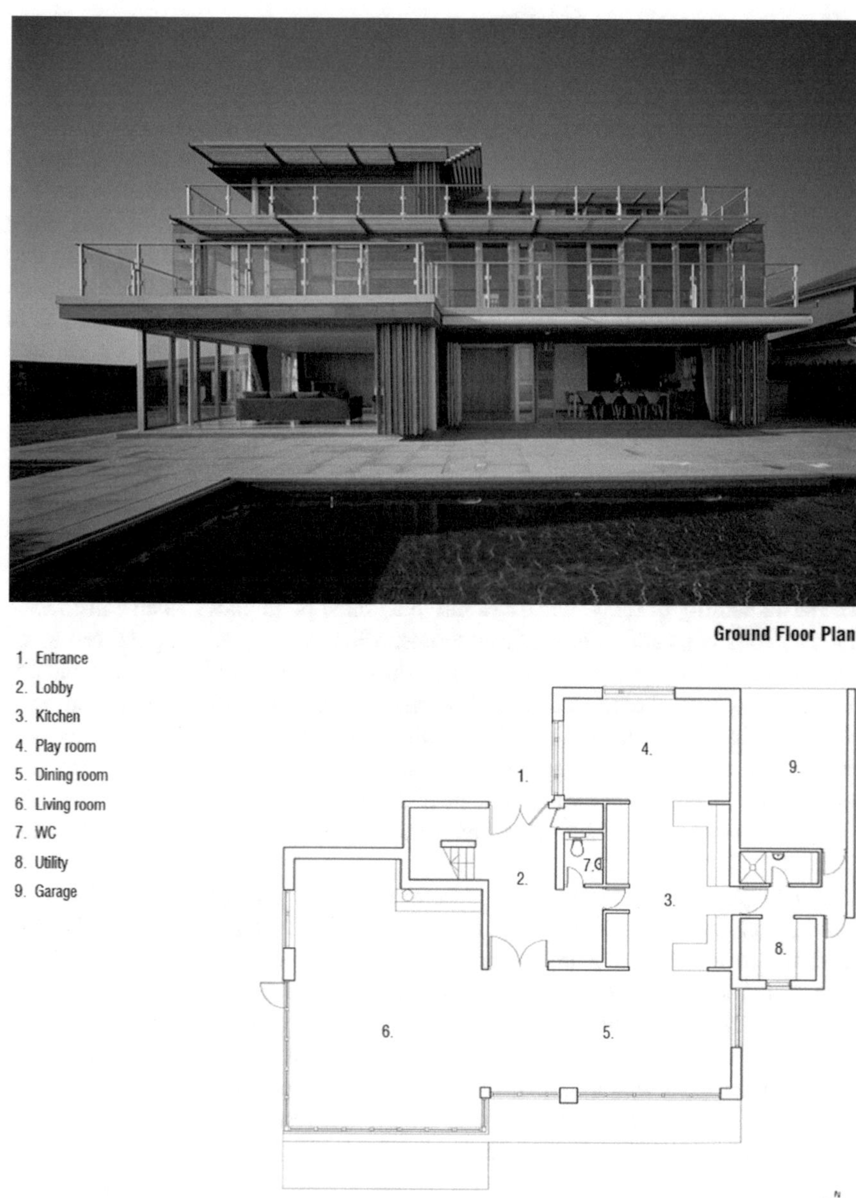

Fig. 2.17 South facing (beach side) elevation of house

of the house, opening the lower interior space out onto the pool area. This also provides natural cross-ventilation, drawing air up through the house and into the bedrooms (Fig. 2.18).

With the timber cladding in mind, we strived in the beach house interior to recreate traditional timber construction techniques, with a workable combination of flowing open plan spaces and more enclosed areas. To create a visual as well as special sense of flow, we used double-sized floor boards, ensuring continuity throughout the levels and areas of the build. The open plan interiors also mean that central areas have the flexibility to be used in multiple ways. We designed our own storage solutions in keeping with this, to ensure that the purpose of the rooms is never defined or limited, except in the bathrooms and kitchen. To cap it off, and again with mind to sustainability, sensitive wiring and heating systems were installed throughout the house to minimize energy usage (Figs. 2.19, 2.20 and 2.21).

2.4 Materials and Details

The sustainability of the build comes into play most particularly when considering the decisions we made in terms of materials. Much of the site's original resources were retained and conserved throughout the duration of the building process, including materials from the demolition and secondary aggregates which were then recycled back into the scheme, ensuring minimal off-site disposal and so increased overall sustainability. Existing foundations, ground slabs, load-bearing masonry walls and elements of the original timber and steel structure were also carefully preserved, then enhanced and woven back into the new overall structure of the build (Figs. 2.22, 2.23 and 2.24).

In choosing materials that were not already on site, we were careful to select those both appropriate to the aesthetic we wished to create and also to the house's existing surroundings. The three-storey structure is therefore clad entirely in iroko timber: iroko not only has the benefit of creating a clean, crisp, modernist finish but also weathers beautifully. In the harsh, salty and blustery seaside environment, the timber forms a silver white patina, mirroring the natural colouring of the shingle on the beach below, resembling, perhaps, driftwood or rocky debris; the house, then, shifts in appearance like an organic product of the landscape that encloses it.

We were also careful to pay minute attention to how the materials used in the build were detailed – take, for instance, the regular pattern of hand-'plugged' concealed screw fixings evident across the entirety of the external timber cladding; a subtle nod in the direction of boatbuilding crafts is relevant not only in the context of the coastal location but also in light of techniques used by the architects who inspired the project, most particularly Paul Rudolph. In line with this thematic reference, and looking more closely at naval architecture and the durability of materials commonly used on seafaring structures, we specified anodised aluminium throughout for all external metal work which would be directly exposed to the elements and the harsh sea climate. So window frames, glass balustrade supports, handrails, cop-

First Floor

1. Bedroom
2. Bathroom
3. WC | shower
4. Void
5. External deck

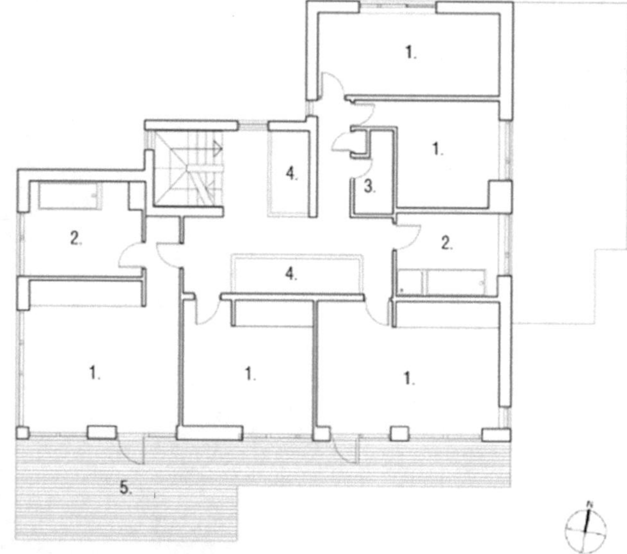

Fig. 2.18 View due West of ground floor external terrace below deep cantileverd balcony eaves

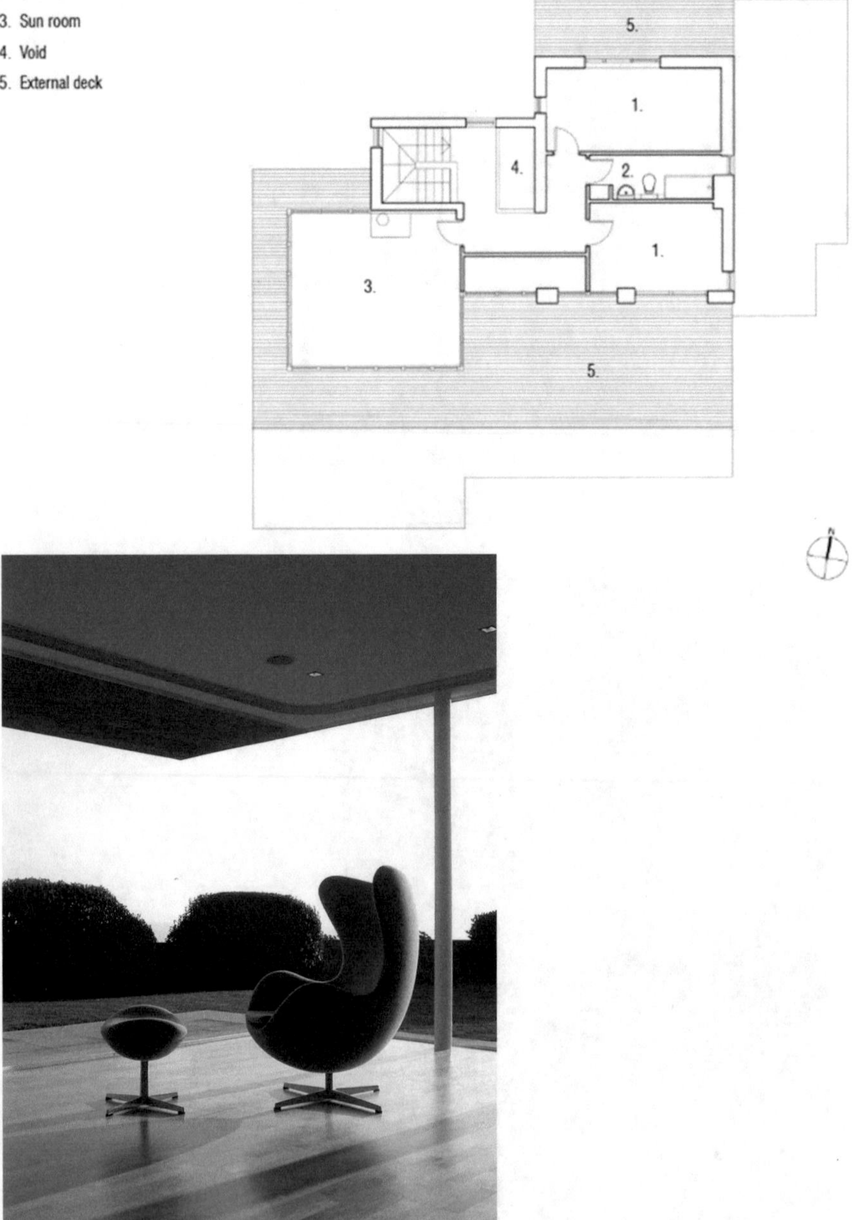

Fig. 2.19 Interior view from the ground floor living room looking out towards the South-West facing corner

Interior views 01 02 03

1. Second floor landing with balcony and sea behind
2. Staircase
3. Staircase handrail detail

Fig. 2.20 The captions are called out next to the image numbers?

Interior views 04 05 06

4. Second floor landing with field behind
5. Staircase rooflight viewed through void from ground floor
6. Staircase rooflight and window detail

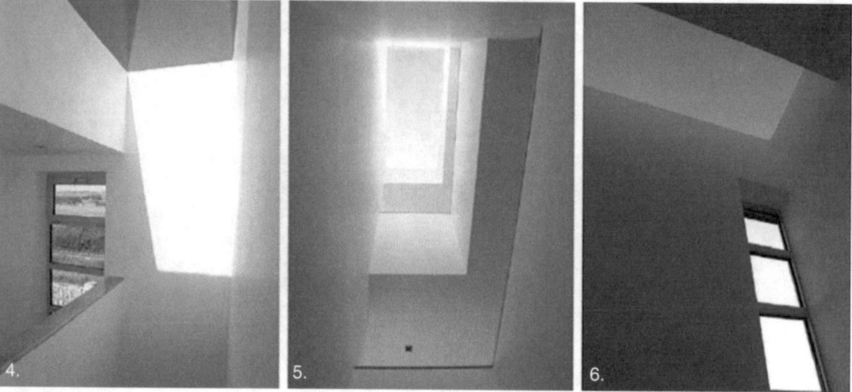

Fig. 2.21 The captions are called out next to the image numbers?

ings, flashings, cappings and finally the south-facing 'brise soleil', providing much needed shade during sunnier periods, were all specified as silver anodised, protecting the metal from oxidization and the formation of a surface patina. And pond lining techniques were specified for the new flat roofs and elevated terraces, being both tough and durable, whilst retaining exceptional flexibility and water proofing.

Fig. 2.22 Interior views of the toplit staircase void

Fig. 2.23 View illustrating the louvered 'brie soleil' taken towards the South-West corner

Fig. 2.24 Exterior detail – Iroko timber cladding

2.5 Challenges

The main challenges of the project revolved around the bureaucratic resistance the build faced at planning stage. Reluctant to include a contemporary property on the site, especially given the nature of the houses which already existed along this part of the coastline, the board refused initially to grant permission, and it was only under appeal that we were able, finally, to move forward with the design. One of our main points of leverage during these discussions was the sustainability of the property and its retention of the original 1970s house within its completed structure – this, alongside the carefully chosen building materials which would mirror the natural colouring of the site's surroundings, provided a persuasive case for the build's continuity with its locality. In conjunction with this, Bexhill already boasts a stunning example of modernist architecture in the form of the De La Warr Pavilion, designed by Russian architects Mendelsohn and Chermayeff; our own site's proximity to the pavilion meant we were able to make a persuasive case for our design's contemporary appearance (Figs. 2.25, 2.26 and 2.27).

2.6 Conclusion

By building Herbrand Walk, we sought to facilitate a way of living that brought the outside in and took the inside out, a smooth-flowing space where architecture and the natural location could flourish cohesively. Decades of experimental domestic design between the 1930s and the 1960s, which progressed so extensively in America, saw little recognition and development here in the UK, in no small part

Fig. 2.25 De La Warr Pavilion, 1935

Fig. 2.26 View of the house at sunset

Fig. 2.27 View of the swimming pool

due to the harsh Northern European climate and growing traditionalist backlash against the perceived ills of the modernist approach. In bringing a touch of sleek mid-century Californian style to a very British location, we hoped to evoke the sentiments of that period which remain largely lacking from England's own architectural history. As our client's so aptly put it, 'It's what we wanted, our Malibu beach house'.

Chapter 3
Natural Ventilation in Hot Seaside Urban Environments

Khaled A. Al-Sallal and Amira R. AbouElhamd

3.1 World Urbanization Trends

The urban population of the world has grown rapidly since 1950, from 746 million to 3.9 billion in 2014 (UN World Urbanization Prospects 2014). In 1950, 30% of the world's population was urban. More people live today in urban areas than in rural areas. According to the UN-WUP 2014 report, 54% of the world's population resided in urban areas in 2014; and by 2050, 66% of the world's population was projected to be urban. The most urbanized regions included Northern America (82% living in urban areas in 2014), Latin America and the Caribbean (80%) and Europe (73%). Africa and Asia remained mostly rural, with 40 and 48% of their respective populations living in urban areas. Continuing population growth and urbanization are projected to add 2.5 billion people to the world's urban population by 2050, with nearly 90% of the increase concentrated in Asia and Africa. The world is projected to have 41 megacities (with more than 10 million inhabitants) by 2030. Several decades ago, most of the world's largest urban agglomerations were found in the more developed regions, but today's large cities are concentrated in the Global South. The fastest-growing urban agglomerations are medium-sized cities, and cities with less than one million inhabitants are located in Asia and Africa.

The growth of the industrial societies in many parts of the world has resulted in fast urbanization with a high-consuming, resource-depleting lifestyle. In some parts of the world, such as the Gulf Cooperation Council (GCC) countries, the fast growth of societies supported by rich energy resources, mainly oil and gas, has also resulted in expansive urbanization and reliance on increased levels of fossil fuel supply to support completely new forms of excessively enhanced lifestyles. Most of the world energy is still produced by non-renewable sources of power generation that are major contributors of greenhouse gas (GHG) emission out of which carbon dioxide

K.A. Al-Sallal (✉) • A.R. AbouElhamd
UAE University, Al-Ain, UAE
e-mail: k.sallal@uaeu.ac.ae

A. Sayigh (ed.), *Seaside Building Design: Principles and Practice*, Innovative Renewable Energy, https://doi.org/10.1007/978-3-319-67949-5_3

(CO_2) is crucial. This, in turn, results in high ecological footprints (World Wildlife Foundation 2010). The highest ecological footprints (measured in global hectares per capita, gha/capita) by far are found in countries like the UAE (10.68 gha/capita; this is the world's highest average ecological footprint), Qatar (10.51 gha/capita) and the USA (8.00 gha/capita) (Global Footprint Network 2010). Buildings consume more energy than any other sector. In the USA, the built environment is responsible for about 48% of all energy consumption and greenhouse gas emissions; and around 75% of all the electricity is used just to operate buildings (Architecture 2030, 2011). The estimated value of building energy sector in the USA was revealed by Architecture 2030, based on data from the US Energy Information Administration (EIA). In Europe, energy consumption by buildings accounts for around 20–40% of the total energy consumption.

Kharecha et al. (2010) classified the available strategies that can substantially reduce building sector GHG emissions into three: the planning and design strategies, the building envelope and material and equipment selection and the added technologies. Natural ventilation was listed as one of the planning and design strategies to reduce GHG emissions.

3.1.1 Benefits of Natural Ventilation

Natural ventilation has several benefits with regard to improving the environmental quality of urban areas/buildings, and thus it is very crucial and wise to consider it in the planning of urban areas and buildings (Santamouris 2005; Al-Sallal 2016b). It can contribute to mitigate problems of indoor air quality by lessening the concentration of indoor pollutants, improve thermal comfort conditions in indoor and outdoor spaces and decrease the energy consumption of air-conditioned buildings. The conditions to achieve these benefits can be outlined as follows: the concentration of outdoor pollutants is lower than that of indoor pollutants; the outdoor temperature is within 'comfort' limits or, in the worst case scenario, does not result in thermal stress of people; and natural ventilation does not cause other environmental and social problems (noise, privacy, etc.).

Santamouris (2005) categorized how natural ventilation could contribute to the well-being of a building's inhabitants as a function of people's life standards and energy use into three main categories:

1. *Decreasing indoor air pollution caused by combustion processes in very poor households*: globally there are more than 2 billion people who live in substandard conditions (i.e. with no access to electricity and modern fuels) who can benefit from improvement of indoor air pollution by natural ventilation.
2. *Improving indoor air quality and indoor thermal conditions for people of low and medium income*: the people in this category (around three billion people) can benefit from natural ventilation since they live in poorly designed buildings usually without cooling equipment, and thus they suffer from high indoor

temperatures during the summer. This would require integration of efficient natural ventilation systems and components into the building design so that it can assist in improving indoor thermal comfort such as wind and solar towers.

3. *Improving thermal comfort, decreasing the need for air-conditioning and improving indoor air quality*: the people in this category live mostly in places that rely on cooling equipment—integration of natural ventilation into the building design and cooling equipment is the main approach in this category. Generally, one should use daytime ventilation technique in mild climates and night ventilation techniques in hot climates (Santamouris 2012). Experimental studies (Santamouris and Asimakopoulos 1987) showed that effective night ventilation in office buildings can reduce the cooling load by at least 55%.

3.1.2 Limitations of Natural Ventilation

Natural ventilation can experience serious reduction of wind speed in dense urban environments. In a previous experimental investigation applied to single-sided and cross-ventilated buildings in ten urban canyons in Athens (Santamouris 2001), airflow rate went down by 90%. Efficient integration of natural and night ventilation techniques can help to improve wind speed in dense urban areas; yet this solution would require full knowledge of wind characteristics, as well as adaptation of ventilation components to local conditions.

Another serious limitation for natural ventilation in urban areas is outdoor air pollution. It was estimated that 70–80% of the European cities that had more than 500,000 inhabitants exceeded WHO annual standard levels regarding one or more pollutants, as reported by Wackernagel et al. (1999). Cleaning of air and filtration of its pollutants are possible only when flow-controlled natural ventilation components are used.

A third limitation is that effective airflow for natural ventilation might produce unacceptable noise levels. Wackernagel et al. (1999) found that more than 65 dB(A) level affect between 10 and 20% of urban inhabitants in most European cities. The OECD (Geros 2000) reported that 130 million people in OECD countries are exposed to noise levels that are unacceptable.

3.2 Current Approaches

In hot climates, a compact urban design with very deep canyons is preferable with regard to shading. However, if there is a cold season, the urban design should include some wider streets or open spaces or both to provide solar access (Johansson 2006). High levels of humidity, which can create thermal discomfort, can be experienced in seaside cities during the times when the wind velocity is low. The following section is a review of the current approaches that investigated how design variables (such as the urban morphology, the greenery and shading and the geometry of the building patterns) affect natural ventilation and the potential of thermal comfort and pollution mitigation. The scope of this review is limited to seaside urban environments in hot climates.

3.2.1 Shading/Greenery and Relation to Urban Canyons

Using passive outdoor shading techniques such as plants and greenery has been widely used as one of the strategies that improve the outdoor air quality. Several studies have investigated how greenery in different urban canyons could improve human thermal environment. A study by Jamei and Rajagopalan (2017) used both onsite measurements and a 3D microclimatic modelling tool ENVI-met to evaluate the outdoor human thermal environment for the existing and future scenarios presented in 'Plan Melbourne' for extreme hot summer days. ENVI-met used satellite images to assist in locating different buildings, vegetation and soil characteristics in the study area. Field measurements were also conducted to validate ENVI-met and to explore additional scenarios. In the field measurements, a portable weather station (3 m above the ground), four HOBO onset data loggers (2 m above the ground) and two comfort carts were used to evaluate seven points across a range of urban canyons. The field measurements were conducted on January 5, 2015, from 7:00 am to 7:00 pm local time under clear sky conditions. The variables studied for the structural plans were increasing building height, adding tree canopy coverage and adding green roofs. The analysis of the results shows that increasing building heights improved the physiological equivalent temperature (PET) by 1–4 °C. Similarly, increasing the aspect ratio was found to better the thermal comfort for both E-W streets and N-S street orientations. The aspect ratio was defined as the average height (H) of the canyon walls and the canyon width (W). On the other hand, while adding a green roof did not show any improvement on PET at pedestrian level, increasing tree canyons coverage area decreased the received mean radiant temperature by 2 °C. Furthermore, integrating public realms with small urban parks leads to 1 °C reduction in the average mean radiant temperature across the study area, and additional cooling effect is noticed from urban parks with 80% tree coverage. The study also found that while increasing the tree canopy coverage from 14 to 40% along with increased building height scenario generated a small reduction on the average mean radiant temperature, the simulation results for long-term planning strategies revealed that increasing the tree canopy coverage from 40 to 50% would lead to a more effective cooling effect from the vegetation.

Another study that was completed by Lai et al. (2014) investigated outdoor thermal comfort under different climate conditions in Northern China using microclimatic monitoring and subject interviews at a park in Tianjin, China. This investigation conducted 11 field surveys that involved microclimatic monitoring, questionnaires, and activity recording between 10:00 and 16:00. These surveys were completed between March 13, 2012, and January 8, 2013. The study revealed that people's preference for additional solar radiation, wind speed and relative humidity was strongly linked to air temperature. It was found that the higher the air temperature, the higher the wind speed and the lower the solar radiation and relative humidity preferred by the occupants, and vice versa.

Makaremi et al. (2012) also investigated the outdoor thermal comfort conditions in hot and humid tropical climate of Malaysia in outdoor shaded areas. The study evaluated the conditions of outdoor spaces based on the major climatic measurement which are ambient air temperature, relative humidity, wind speed and mean radiant temperature. Data loggers, probe with omnidirectional hot wire and globe thermometer probe were used to evaluate these climatic measurements. Additionally, the study captured the thermal perception of subjects by distributing random surveys on both international and local student in the campus of Universiti Putra Malaysia. The two studying areas in the university campus, which contain different environmental conditions, were investigated with each study area encompasses two study locations: space I and space II. The field investigations of this study were conducted on March 16–17 and April 7–8, 2010, from 9:00 am to 5:00 pm. The physiological equivalent temperature (PET) thermal comfort index was utilized to assess the thermal comfort conditions. It was found that PET in the selected shaded outdoor spaces of the campus was higher than the comfort range defined for tropical climate (PET < 30 °C). Nevertheless, these acceptable conditions (PET < 34 °C) normally occurred during the early hours of measurement (9–10 am) and late afternoon (4–5 pm). Additionally, it was found that the locations with high level of shading obtained from plants and surrounding buildings had a longer thermal acceptable period for students. The study concluded by highlighting the substantial role of environmental factors such as wind speed on human thermal comfort level in outdoor spaces.

3.3 Walkway Orientation

Walkway orientation is another crucial factor to be considered when designing for pedestrian thermal comfort. A study by Rismanian et al. (2016) in the hot and dry climate in the traditional city of Sirjan, Iran, has investigated the effects of walkway orientation on the airflow rate created by its free movement. Airflow in the walkway was simulated by the commercial CFD code FLUENT 6.3.26 on June 21, at 4 pm, which considers as the hottest time of the year. The geometry of the simulated walkway was 12 m length, 2 m width and 3 m height. The study modelled the entire region surrounding the walkway, and computational grids were generated using the Gambit package. Then, temperature, velocity distribution, maximum velocity and produced natural airflow in the walkway were measured to investigate the natural ventilation rate at seven different angles and at 15° intervals. The main outcome of this study is that the best walkway orientation that generates the maximum air ventilation in the walkway was when it makes an angle of –50° or +40° relative to a northerly direction. This higher airflow rates in hot and dry climates improve pedestrian feeling of comfort as air will evaporate swat and thus improves people feeling of comfort.

3.4 Urban Morphology

Urban geometry hugely affects wind speed and hence the thermal comfort at pedestrian level. In a review by Yannas (2001), he highlighted that wind velocities in cities are generally lower than those in the open country because of the sheltering effect of buildings. This results in a lower rate of heat dissipation by convection. Another issue is that tall buildings and the channelling effects of street canyons lead to complex airflow patterns and produce turbulence. In addition to that, most cities lack substantial green areas and bodies of water and thus obtain little benefit from evaporative cooling. Yannas also discussed that the built form influences the magnitude of the air movement inside and outside the buildings. The following subsections discuss some studies that highlight the most significant geometrical aspect when designing for pedestrian thermal comfort.

Plan Area and Building Aspect Ratios The plan aspect ratio and the building aspect ratio are the most two crucial geometric parameters affecting the wind environment at pedestrian level. Abd Razak et al. (2015) investigated the performance of pedestrian wind at four major cities in Klang Valley. Field measurement was conducted in the selected locations, where the mean wind speed at pedestrian level, wind direction and mean wind speed above the canopy layer were measured using anemometers (WatchDog 2550 Weather Station) for duration of 3 h with 1 min interval. The anemometers were placed 2 m above the rooftop of the tallest building in each case to find the wind direction and mean wind speed. Then the pedestrian mean wind speed was measured at 1.5 m from the ground level. A square area of 745 m × 745 m that consists of high-rise building and low-rise building was selected to be the study area in each case. The mean wind speed ratio was plotted against frontal area and plan area ratio. The plan area ratio was defined as the ratio of projected building roof area to the ground surface area. On the other hand, the frontal area ratio was defined as the ratio of building surface area normal to the upwind to the ground surface area. The results of the study showed that air velocity at pedestrian level was greatly affected by the urban morphology. High-density cities were found to produce low mean wind speed and cause higher pollutant concentration at pedestrian level. Additionally, the mean wind speed was found to dramatically decrease with the increase of plan area ratio, and it exponentially decreased with the increase of frontal area ratio. In addition, it was concluded that the frontal area would provide a better parameter to evaluate the performance of urban ventilation.

Plan Area Ratio, Block Aspect Ratio and Height Variability Understanding pedestrian wind environment was also the aim in an earlier study by Abd Razak et al. (2013) who investigated the effect of building height variability on pedestrian wind environment under an idealize urban model using large eddy simulation (LES). The study simulated five types of idealized uniform staggered block arrays with different aspect ratios and other non-uniform arrays consisting of two types of blocks with different height. All blocks had a square base of 25 m, and the simulation was done over a unit area which consists of four blocks. The study used a wall function based on logarithmic wind profile to be imposed on the bottom boundaries and surfaces of blocks, and a free-slip condition of the Neumann type was used at the top of the domain. Then, the spatially averaged mean wind profiles were analysed for various conditions of plan area ratio, block aspect ratio and height variability. Block aspect ratio is defined as the ratio of roof area to frontal area of a block. The simulated results revealed that the pedestrian wind speed depends not only in plan area ratio but also in building aspect ratio; therefore, the frontal area ratio which is the product of these two is the most important parameter in estimating the pedestrian wind speed. The study also derived a simple exponential equation for predicting the pedestrian wind speed as a function of the frontal area ratio, which is applicable to various building aspect ratios and amounts of height variability.

3.5 Building Patterns and Orientations

Building patterns can also hugely influence the wind environment at pedestrian level. Several studies explored the influence of different building patterns arrangement on the wind environment. Hong and Lin (2015), for example, investigated the effects of building layout patterns and tree arrangement on the outdoor wind environment and thermal comfort at the pedestrian level in Beijing. They used numerical simulations using simulation platform for outdoor thermal environment (SPOTE) that consist of an air model, a vegetation model, an underlying surface model and a general radiation calculation model. Furthermore, calculation of radiation, convection, conduction and airflow was carried out considering the vegetation influences. The study analysed six building configurations; the first configuration consists of two parallel rows of housing blocks. The second and third configurations represent a T-shaped central space with, respectively, upstream and downstream opened side. The fourth configuration represents a staggered pattern, while the fifth configuration is characterized by a square central space. The last configuration is divided into four zones. In addition to that, trees are planted along the pedestrian road around each building were created in the simulation tool to represent the real-case scenario in Beijing. The area of the site was 80 m × 80 m, and the measurement of each housing block was 30 m × 15 m × 18 m with rises for six storeys. It was concluded depending on the simulated results that the parallel arrangement of building to the prevailing wind direction could accelerate horizontal vortex airflow at the edges. Such airflow could strengthen the convective exchange efficiency of hot air in low altitude and cold air in high altitude. It can also obtain pleasant wind environment and thermal comfort at pedestrian level. Moreover, it was observed that configurations with a square central space articulated by buildings and oriented towards the prevailing wind can offer better exposure to air currents and experience better air movement.

The research also recommended configuration 3 in the case of 0° wind direction for better wind-driven ventilation and comfortable thermal environment resulting from the maximum wind pressure difference.

The wind environment at pedestrian level was also the concern of Kubota and Ahmad (2005) who investigated the average wind velocity on selected residential areas in Johor Bahru metropolitan city. The study investigated three residential neighbourhoods with different housing patterns based on their housing density; and five areas out of them were chosen for wind tunnel tests. The wind tunnel was 1.4 m × 1.4 m and 9 m in length and all the models in the test were scaled in 1:500. With the focus to evaluate different housing patterns, 49 measuring points were placed equally outside of the buildings in each area. Both wind direction and wind velocity at 1.5 m height of each point were then measured in 16 wind directions. The experimental results showed that the perpendicular arrangements of terraced houses to the wind direction are not recommended as this arrangement was found to reduce the average wind flow. It was also noticed that wind was blown through the spaces between semi-detached houses from windward side to leeward side; therefore, semi-detached houses are recommended in order to avoid the increase of weak wind flow areas. Furthermore, the relationship between the gross building coverage ratio and the mean wind velocity ratio of the cases studied in Malaysia was compared with that studied previously in Japan. It was found that the mean values of wind velocity ratio of the Malaysian cases were slightly higher than that of Japan. This was referred to the straighter and longer terraced houses occurred in Malaysia.

3.6 Design Variation of the Building Blocks/Canyons

The existence of heat island in Muar, Malaysia, was investigated (Rajagopalan et al. 2014). The study examined the effects of different urban geometries on the wind flow. The research used fixed station measurements in July 2011 where three measurements were carried out during the day: 2.30–3.30 pm, 7.30–8.30 pm and 11.30 pm to 12.30 am. The study used HOBO onset data loggers inserted inside a white colour cylindrical tube in order to be protected from the sun and possible precipitation. The measurements were conducted at the height of 750 mm from the ground. In addition to the field measurements, numerical simulations were also conducted using IES virtual environment in an area that had a radius of 500 m within the city centre to examine the wind pattern in Muar City. The study area consists of four rows of buildings and three main canyons with most of the canyons are asymmetrical with the average height of two to three storeys. Various scenarios were tested by the simulation tool. The first scenario reduced the height of the high-rise building to match the average height of low-rise buildings, and thus the entire study area had the same height. In the second scenario, an opening was cut through the middle of the tall building to encourage wind penetrating through it. The third scenario added more high-rise buildings to match the future development of the city. And the last scenario created a step-up configuration to enhance wind flow. The step-up configuration was defined as an urban configuration where the height of the upwind building is less than the height of the downwind building. In each scenario, the wind speeds were measured using a handheld Kanomax hotwire anemometer for five locations within 1 h and then repeated in the morning, afternoon and night. It was found that in Muar City, the combination of tall buildings and narrow streets traps heat and reduces the airflow, resulting in high temperatures. The study also highlighted the significance of building height in the process of urban planning as they have a great impact on the airflow at pedestrian level. It was also found that for areas with low wind velocity, vegetation and shading could create thermally comfortable public spaces. This greenery arrangement was found the best effective improvement for the overall natural ventilation as it can distribute the wind evenly and allow the wind to reach the leeward side of each building. This improved the overall natural ventilation and thermal comfort at pedestrian level.

Fan et al. (2017) evaluated both the air quality and wind comfort induced by building openings at the pedestrian level of street canyons. The study predicted air pollutant concentrations and wind velocities induced by building openings by employing a series of CFD simulations using ANSYS Fluent software based on standard k-ε model. Three scenarios were modelled and studied to compute the street air-wind index values for both isolated and non-isolated canyons. The first scenario consists of two identical building blocks without or with openings that encompass street canyon with 18 m width and 180 m length with three aspect ratios (AR2, AR4 and AR6). The aspect ratio is defined as building height over street width. Six types of models with different opening shapes at different locations were investigated. These are as follows: (1) separation at middle part, (2) separations at both sides, (3) voids at ground level, (4) voids at both ground and podium level, (5) permeable elements at ground level and (6) permeable elements at podium level. Both wind velocity and pollutant concentration were extracted at 1 m interval and been used to compute the proposed street air-wind index values. This street air-wind index helped evaluating the effect of building openings on air quality and wind comfort at the pedestrian levels for both isolated and non-isolated canyon configurations. The most important conclusions of this study were that permeability values of 10% were shown to be adequate for improving the pedestrian environment inside street canyons and increasing this value to 20% could bring an additional increase of 20% in street air-wind index values in AR2, 16% in AR4 and 11% in AR6. Additionally, it was found that permeable elements at the ground level were the most effective strategy for AR2 isolated canyons, while permeable elements at the podium level were found to be most effective for AR4 and AR6 isolated canyons. As for non-isolated canyons, it was recommended to place permeable elements at the ground level of buildings at both sides of target canyon and permeable elements at the podium level of buildings located in adjacent canyons. Another outcome of this research was that the effect of openings on the street air-wind index was found to be different between isolated and non-isolated canyons. In isolated canyons, the increase in the street air-wind index value was mainly because of the improvement in air quality but not wind comfort. On the contrary, openings were not always effective for non-isolated canyons if there were pollutant sources in adjacent street canyons. Introducing openings to adjacent canyons along with openings to the target canyons was recommended.

3.7 Case Study: Dubai, UAE

The UAE lies between 22° 50′and 26° north latitude and between 51°and 56° 25′ east longitude. This unique location on the Tropic of Cancer with land-sea distribution provides its region a subtropical desert climate, predominantly hot dry with narrow coastal ribbon that has hot humid climate. The city of Dubai is located on this narrow coastal area with latitude 25° 15′ north, longitude 55° 20′ east and elevation of 8 m above sea level. Most days in Dubai are sunny throughout the year, with extremely hot and humid summers and warm winters. The summer average high is around 41 °C (106 °F) with overnight lows around 30 °C (86 °F) and can go as high as 48 °C in the summer (Wikipedia 2017). The winter average high is 23 °C (73 °F) with overnight lows around 14 °C (57 °F). The minimum and maximum wind speed in the month of December would be 2.98 m/s and 6.16 m/s, respectively, whereas in the month of June, it would be 3.75 m/s and 7 m/s, respectively. Light rainfall and fog are characteristics of the winter season, with high relative humidity in summer (80–90%). The author investigated earlier the thermal comfort issue in Dubai (Al-Sallal and Al-Rais 2011) where the annual weather conditions of Dubai were plotted on the psychrometric chart using Climate Consultant 4.0 (Milne). The analysis showed that of the plotted points (hours of the year), 48% exceeded the comfort range (20–27.5 °C) or overheated period, 33% was a comfortable period and 17% was an underheated period.

The passive cooling performance in both the traditional and the modern urban contexts in the hot humid climate of the city of Dubai was investigated using computational fluid dynamics (CFD) modelling (Al-Sallal and Al-Rais 2011; Al-Sallal and Al-Rais 2012). In the traditional urban context investigation, three cases were simulated in both laminar and turbulence air movements for Al Ras area. This district is located at the heart of the old city of Dubai (Fig. 3.1). The famous traditional merchandise markets of Dubai including the gold and the spice markets predominantly occupy this region. This area is located at the heart of the old city of Dubai with average area of 23,156 m^2. The aspect ratio of the street canyons within the studied area ranges from 0.44 to 2, and the population density per squared kilometre in this area was 12,828–22,313. A laminar case was conducted first followed by two turbulence modelling cases conducted for both summer and winter. The results of the study revealed that the traditional urban context in Dubai resists funnelling of airflow inside the narrow streets especially when the prevailing wind speed is less than 3 m/s (Fig. 3.2). However, when the wind velocities inside the streets were checked against the Beaufort scale, most locations were found to be comfortable in both winter and summer. Additionally, when wind speed increases (>5 m/s), it can provide better potential for thermal comfort as it reaches deeper inside the traditional narrow streets. Another finding is that in most locations (49–57% of the studied area) that have street canyon aspect ratio AR = 2–0.67, wind speeds range from light

Fig. 3.1 Aerial view (Source: Google Earth) and street-level photos of the street canyons and open spaces in the traditional urban context of Al-Ras district, Dubai, UAE

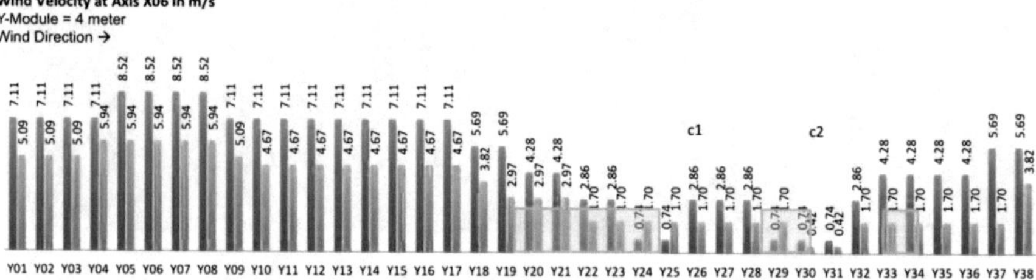

Fig. 3.2 Wind velocity profile of the both winter and summer results at axis X06 in the traditional urban context of Al-Ras district, Dubai, UAE

breeze to gentle breeze (according to Beaufort scale), which has the potential to provide natural cooling with around 5–8.5 °C (i.e. a lower-temperature comfort sensation with basic assumption of 1.3 MET and 0.4 CLO of summer clothing). Narrow street canyons (4 m and less) were also found to accelerate wind speed passing through it, hence resulting in a better passive cooling performance; however, these streets sometimes create eddies especially when they include many bending angles.

In the modern urban context investigation, the same approach was implemented to simulate Al Mankhool modern district in Dubai. This area is very vibrant and well known as a modern commercial district (Fig. 3.3). It also includes the famous BurJuman shopping mall. The population density per squared kilometres in this zone of Dubai is 7807.654–12,828.014. The area of the simulated site is approximately 374.31 m in width and 532.17 m in depth, with eight floors as average building height (except two buildings). It included mainly two types of street canyons with aspect ratio equal to 2.8 and 1.7, respectively, and some open spaces with

Fig. 3.3 Aerial view (Source: Google Earth) and street-level photos of the street canyons and open spaces in the modern urban context of Al-Mankhool district, Dubai, UAE

AR = 0.63–1.56. It was found that the wind flow decreased when hitting buildings, funnelled by the wider street canyons, and then increased once again when going to the free stream (Fig. 3.4). Additionally, wind flow was noticed to funnel randomly between buildings. While this trend improved wind flow circulation in some places and hence maximized thermal comfort, it created vortices at building corners causing discomfort to some extent at the pedestrian level. Another remarkable finding is that wider street canyons with an aspect ratio (AR) = 1.75 created more comfortable wind velocity than those created in narrow street canyons. Furthermore, whereas the wind velocities stayed stable with limited fluctuation in longer street canyons in summer and winter, wind velocities were found increasing substantially in open spaces such as parking areas in both summer and winter.

Another research on the city of Dubai investigated the influence of moves within different thermal environment on pedestrians' thermal sensation aiming at improving their comfort levels (Al Sabbagh et al. 2016). This study evaluated a series of

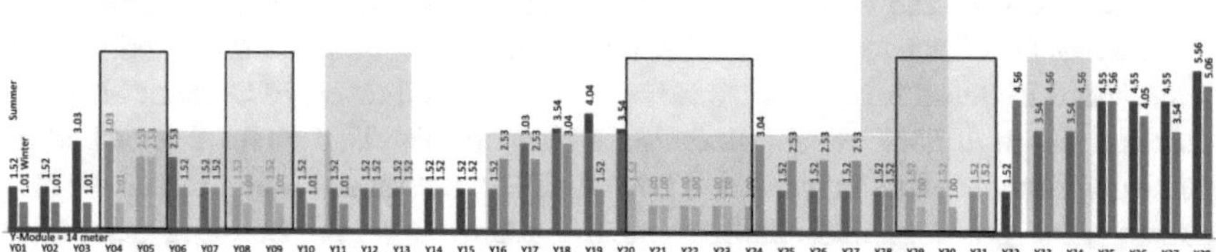

Fig. 3.4 Wind velocity profile of the both winter and summer results at axis X10 in the modern urban context of Al-Mankhool district, Dubai, UAE

short walks using subjective assessments of thermal sensations votes (TSV) and thermal comfort votes (TCV) that were collected from six subjects (three males and three females) to assess their tolerance to walking in the sun in January, May and November at 13:20–13:40. Subjects were 32–35 years old with a body mass index of 26–29 and clo value of 0.7. A typical sunny outdoor path was selected adjacent to a building exit in Jumeirah Lakes Towers (JLT) district that has a typical light-grey concrete tiles and had no vegetation. Each subject was asked to leave the air-conditioned building and walk along the path for 15 min, where the TSV and TCV were recorded every minute during the walk using a Testo 410-2 handheld instrument. Then, subjects were asked to walk in a shaded area for 4 min where an air velocity of 3.2 m/s was directed towards their face and upper body using a fan, and then they were asked to move back again to the sun for 1 min. In addition to that, the skin temperature and sweat rate were observed and measured during the whole experiment. The results showed that the large temperature differences between indoor and outdoor environments affected thermal comfort, whereas the presence of wind and shade assist improves thermal sensation and comfort and thus increases walkability. It was also found that the minimum tolerable walking time without shade was 8 min during the hot period of the year and 10 min during the warm and mild periods. Additionally, the warm period during November was the best to improve thermal comfort and thermal sensation by promoting solar control and wind movement. It was also found that in Dubai, where high temperature and high humidity levels excite, the wind had higher potential in improving pedestrian comfort and sensation of moist skin.

3.8 Conclusions

Rapid urbanization puts a huge burden on urban buildings and its overall energy consumption. This is evidently documented by the United Nations (UN) recent reports that showed 54% of the world's population resided in urban areas in 2014; and by 2050, 66% of the world's population was projected to be urban. Thus, ventilation, in particular, natural ventilation, is one of the means that will help significantly in mitigating air pollution and improving air quality in the urban environments, in improving thermal comfort and in reducing buildings' energy consumption on both the architectural and urban scales. This chapter discussed that effective airflow for natural ventilation might face three limitations: the reduction in wind speed in dense urban environments, the increased levels of outdoor air pollution and the high noise levels produced by the airflow. This chapter reviewed the current approaches that investigated how design variables including the urban morphology, the greenery and shading and the geometry of the building patterns in seaside urban environments affect natural ventilation and discussed their potential of providing thermal comfort and pollution mitigation. Finally, as a case study, Dubai was analysed with regard to passive cooling and thermal comfort in three researches. Two of which covered both the traditional and the modern urban environments in order to give a full range of different design variables and its impact on airflow and passive cooling. The third one analysed the influence of moves on pedestrian thermal sensation.

References

Abd Razak, A., Hagishima, A., Ikegaya, N., & Tanimoto, J. (2013). Analysis of airflow over building arrays for assessment of urban wind environment. *Building and Environment, 59*, 56–65.

Abd Razak, A., Rodzi, M. A. M., Jumali, A. H., & Zaki, S. A. (2015). Analysis of pedestrian-level wind velocity in four neighbourhoods in Klang Valley. *Jurnal Teknologi, 76*, 25–29.

Al-Sallal, K. A., & Al-Rais, L. (2011). Outdoor airflow analysis and potential for passive cooling in the traditional urban context of Dubai. *Renewable Energy, 36*, 2494–2501.

Al-Sallal, K. A., & Al-Rais, L. (2012). Outdoor airflow analysis and potential for passive cooling in the modern urban context of Dubai. *Renewable Energy, 38*, 40–49.

Al-Sallal, K. A. (2016a). *Low Lnergy Low Carbon Architecture: Recent Advances and Future Directions*, a book in the book series "*Sustainable energy developments*". CRC Taylor and Francis Group.

Al-Sallal, K. A. (2016b). Passive and Low Energy Cooling. A chapter in K. Al-Sallal (Ed.), *Low Energy Low Carbon Architecture: Recent Advances & Future Directions*, (pp. 17–62). CRC Taylor and Francis Group.

Al Sabbagh, N., Yannas, S., & Cadima, P. (2016). Improving pedestrian thermal sensation in Dubai. In *PLEA 2016 – 36th International Conference on Passive and Low Energy Architecture, Cities, Buildings, People: Towards Regenerative Environments,* Los Angeles.

Architecture 2030. (2011). *Architecture 2030* [Online]. http://architecture2030.org/. Accessed 1 Nov 2013.

Fan, M., Chau, C., Chan, E., & Jia, J. (2017). A decision support tool for evaluating the air quality and wind comfort induced by different opening configurations for buildings in canyons. *Science of the Total Environment, 574*, 569–582.

Geros, V. (2000). *Ventilation nocturne: Contribution a la reponse thermique des batiments.* Thèse, INSA, Lyon.

Givoni, B., Noguchi, M., Saaroni, H., Pochter, O., Yaacov, Y., Feller, N., & Becker, S. (2003). Outdoor comfort research issues. *Energy and Buildings, 35*, 77–86.

Global Footprint Network. (2010). *Ecological footprint atlas 2010.* http://www.footprintnetwork.org. Abgerufen am 4 Feb 2011.

Hong, B., & Lin, B. (2015). Numerical studies of the outdoor wind environment and thermal comfort at pedestrian level in housing blocks with different building layout patterns and trees arrangement. *Renewable Energy, 73*, 18–27.

Jamei, E., & Rajagopalan, P. (2017). Urban development and pedestrian thermal comfort in Melbourne. *Solar Energy, 144*, 681–698.

Johansson, E. (2006). Influence of urban geometry on outdoor thermal comfort in a hot dry climate: A study in Fez, Morocco. *Building and Environment, 41*, 1326–1338.

Kazmerski, L., Gallo, C., Sala, M., & Sayigh, A. (1998). *Architecture: Comfort and energy*. Oxford, UK: Elsevier Science.

Kharecha, P. A., Kutscher, C. F., Hansen, J. E., & Mazria, E. (2010). Options for near-term phase-out of CO2 emissions from coal use in the United States. *Environmental Science & Technology, 44*, 4050–4062.

Kubota, T., & Ahmad, S. (2005). Analysis of wind flow in residential areas of Johor Bahru City. *Journal of Asian Architecture and Building Engineering, 4*, 209–216.

Lai, D., Guo, D., Hou, Y., Lin, C., & Chen, Q. (2014). Studies of outdoor thermal comfort in northern China. *Building and Environment, 77*, 110–118.

Makaremi, N., Salleh, E., Jaafar, M. Z., & Ghaffarianhoseini, A. (2012). Thermal comfort conditions of shaded outdoor spaces in hot and humid climate of Malaysia. *Building and Environment, 48*, 7–14.

Milne, M. *Climate Consultant Software*. Los Angeles: Dept. of Architecture and Urban Planning, UCLA. http://www.energy-design-tools.aud.ucla.edu/

Rajagopalan, P., Lim, K. C., & Jamei, E. (2014). Urban heat island and wind flow characteristics of a tropical city. *Solar Energy, 107*, 159–170.

Rismanian, M., Forughi, A., Vesali, F., & Mahmoodabadi, M. (2016). Investigation of the effect of walkway orientation on natural ventilation. *Scientia Iranica. Transaction B, Mechanical Engineering, 23*, 678.

Santamouris, M. (2001). *Energy in the urban built environment*. London: James and James Science.

Santamouris, M. (2005). Energy in the urban built environment: The role of natural ventilation. *Natural ventilation in the urban environment: Assessment and design*, 1–19.

Santamouris, M., & Asimakopoulos, D. (1987). *Passive cooling of buildings*. London: James and James Science.

Santamouris, M. (2012). Energy in the built environment. A chapter in C. Ghiaus & F. Allard (Eds.) *Natural ventilation in the built environment: assessment and design*. London, UK.: Earthscan.

United Nations, Department of Economic and Social Affairs, Population Division (2014). *World Urbanization Prospects: The 2014 Revision, Highlights*. Internet: https://esa.un.org/unpd/wup/publications/files/wup2014-highlights.Pdf, ST/ESA/SER.A/352.

Wackernagel, M., Onisto, L., Bello, P., Linares, A. C., Falfan, I. S. L., Garcia, J. M., Guerrero, A. I. S., & Guerrero, M. G. S. (1999). National natural capital accounting with the ecological footprint concept. *Ecological Economics, 29*, 375–390.

Wikipedia. (2017). *Climate of Dubai* [Online]. Accessed 23 May 2017.

World Wildlife Foundation. (2010). *Living planet report 2010: Biodiversity, biocapacity and development*, Switzerland.

Yannas, S. (2001). Toward more sustainable cities. *Solar Energy, 70*, 281–294.

Chapter 4
Architecture and the Sea, the Situation in the Netherlands

Wim Zeiler

The Netherlands or "low countries" is a very densely populated country of which 60% is vulnerable to flooding, 26% of the Netherlands is below sea level, and its peat-rich agricultural soil is subsiding even as climate change is raising sea levels. Climate change has the potential to increase sea levels by several feet, while the lowest city within the Netherlands lies 7 m (23 ft.) below sea level. They have learned a lot the hard way from floods in the past.

The low-lying Netherlands has been fighting back water for more than 1000 years, when farmers built the first dikes. Since the fourteenth century, windmills have been pumping water off the land to keep the low-lying lands dry. A view at Kinderdijk (see Fig. 4.1) gives a beautiful view on windmills, built around 1740; they symbolize Dutch water management and were declared to be a UNESCO World Heritage.

The Zuyderzee Flood of 1916 was a defining moment in Dutch history. The Zuyderzee Works, a series of dikes to reclaim this shallow inlet, was made in part due to the outcome of this flood. It included a large dike, the Afsluitdijk, which in turn allowed for further reclamation of the surrounding area. Unfortunately, further flooding occurred in 1953 in another part of the Netherlands, with devastating consequences in loss of the lives of 1835 people and ruined productive agricultural lands with the salts of seawater. This led to the famous Delta Works. The Delta Works (Dutch, Deltawerken) is a series of construction projects in the southwest of the Netherlands to protect a large area of land around the Rhine-Meuse-Scheldt delta from the sea. It consists of dams, sluices, dikes, and storm surge barriers to shorten the Dutch coastline, thus reducing the number of dikes that had to be raised. The Delta Works has been declared as one of the Seven Wonders of the Modern World by the American Society of Civil Engineers.

W. Zeiler (✉)
Faculty of the Built Environment, University of Technology Eindhoven, Eindhoven, Netherlands
e-mail: W.Zeiler@tue.nl; w.zeiler@bwk.tue.nl

A. Sayigh (ed.), *Seaside Building Design: Principles and Practice*, Innovative Renewable Energy, https://doi.org/10.1007/978-3-319-67949-5_4

Fig. 4.1 Dutch windmills at Kinderdijk (https://www.kinderdijk.com/area/unesco-world-heritage-windmills-kinderdijk)

Still the Netherlands had to continuously learn from more recent past mistakes – a 1977 report warned about the weakness of the river dikes but was ignored because it involved demolishing houses. It took floods in 1993 and again in 1995, when more than 200,000 people had to be evacuated, to put new plans into action. Stronger and higher river dikes now prevent flooding from water flowing into the country by the major rivers Rhine and Meuse, while a complicated system of drainage ditches, canals, and pumping stations keep the low-lying parts dry for habitation and agriculture.

The Netherlands would not be possible in the future without protection from the unpredictable North Sea. Natural sand dunes and constructed dikes, dams, and floodgates provide defense storm surges from the sea. Without an extensive network of dams, dikes and dunes, the Netherlands would be especially prone to flooding. The North Sea is stronger and more unpredictable than the three Dutch rivers that flow through the country. As a predicted outcome of global climate change, sea level rise could impact the Netherlands drastically, leading to social and economic devastation. The Dutch are extremely proud of their water management, and they have eight million people [almost half the population] living below sea level who depend on it.

The Dutch architecture industry is innovative and is highly influenced by the problems associated with water and the growing concerns for the environment. No longer only the ability of a building to meet only the direct functional performance for the occupants and the organizational requirement within the economical boundary constrictions is of importance but also its sustainably. Its sustainability

Fig. 4.2 Designing: to satisfy a need

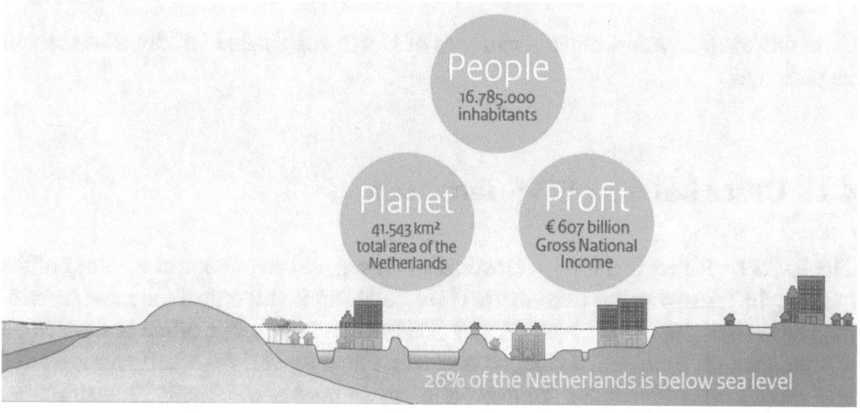

Fig. 4.3 The quantitative representation of the Dutch 3P approach

determines the ecological value and the relationship between building and environment. Sustainability includes many design aspects such as energy and water consumption and the use of materials, emissions, and waste. Above all, it is important to maintain the balance between people, planet, and profit, the 3P approach, for all value frameworks; see Fig. 4.2.

As a result, Dutch architects are active in thinking about the sustainability of their designs and ways to achieve this in their designs. The approach is based on the integration of the different perspective of all disciplines involved within the building design process and to make the abstract ideas not only qualitative but as much as possible also quantitative. See, for example, what the 3P approach means in the Dutch context; see Fig. 4.3.

Fig. 4.4 RWS office Terneuzen

In this chapter, a few recent examples of Dutch sustainable buildings at sea will be presented.

4.1 Office Building RWS Terneuzen

The RWS (= Rijkswaterstaat = Governmental sea and river control agency) office building in Terneuzen has been in use since 2000 and is still one of the most sustainable office buildings in the Netherlands (De Flander 2011). The office is situated on the complex of locks on the Ghent-Terneuzen Canal toward the sea. It was intended as a model office block for "human- and environmentally-friendly" construction. The building's remarkable character and the design that incorporates the users' specific wishes and its surroundings combine to provide the users with an unusual building that sets them apart and a building that is greatly enjoyed by its users (Pötz and Bleuzé 2009).

The building has a triangular shape with an atrium in the middle to utilize maximum daylight; see Fig. 4.4. On top of the atrium, solar PV cells are placed to generate electricity and are positioned to act as sun blinds. Solar thermal panels produce warm water. The reduction of energy demand by artificial lighting and a good visual comfort were the driving forces to create a good daylight situation. For energy conservation during winter and avoidance of solar gains during summer, the glazed area in the facades was limited to 30%, with an overhang above the window designed in relation to the orientation of the facade. Via the atrium, additional daylight is provided through the glazed roof and large windows in the separating walls.

Fig. 4.5 Atrium and glazed roof picture of RWS Terneuzen (Photo credits: OpMAAT)

4.1.1 Energy Concept

Specific attention has been given to temperatures in the atrium. Direct solar incidence through the glazed roof is mainly prevented by the PV cells, which are placed directly above the glazing (facing south); see Fig. 4.5. Moreover, an internal solar shading screen under the roof combined with automatically controlled ventilation valves above this screen reduces solar gains. The absence of the internal screen in the atrium during the first summer period of the building proved the necessity. Design calculations indicated that the ventilation system is important for controlling overheating during summer. Fresh air supply needs a ventilation rate of $n = 1.5$ h-1. The assumption is that above an interior temperature of 23 °C, occupants open the windows. In that case, the ventilation rate is estimated to increase to $n = 6$ h-1, which is necessary to effectively avoid overheating. Conditional night ventilation must be applied.

The building has natural ventilation. A natural ventilation system involves the risk of insufficient performance in case of no wind or no thermal stratification when there are no temperature differences. During the design, all different outside conditions were simulated. The results showed that, due to a good prevalent wind profile on the site at the Dutch coast, during approximately 95% of the time, the natural ventilation concept would function adequately. The simulations indicated that normal inlet grills would result in cross ventilation and an unequal distribution of airflow's throughout the building (van der Aa et al. 2002). The application of electronically, by the building management system (BMS) controlled, constant flow inlet grills makes it possible to constantly adjust the settings of the grills as a function of the air velocity through the opening, which also prevents an overshoot of airflow. During winter, the grills are closed after working hours. The occupants also have the possibility to manually overrule the control system. From the office rooms, the polluted airflow goes to the central atrium through overflow openings in the internal separation walls; see Fig. 4.6. The air is then extracted from the atrium by a large 7-m-high chimney with a 1 m diameter. This was needed to create sufficient under-pressure to extract the air under all weather conditions. The chimney is opened and closed by a controlled inlet side grill.

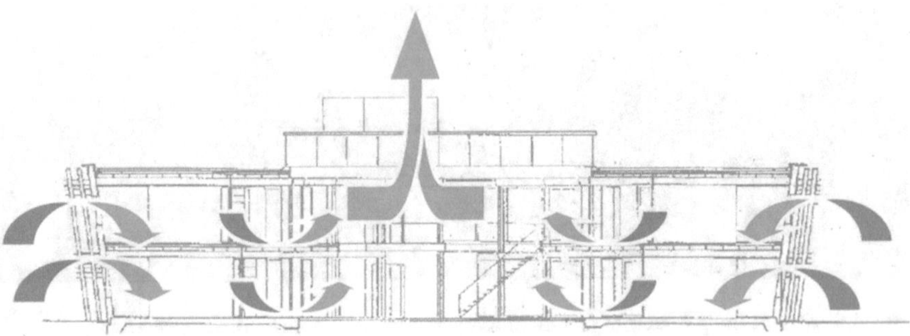

Fig. 4.6 Projected airflow within the RWS Terneuzen building

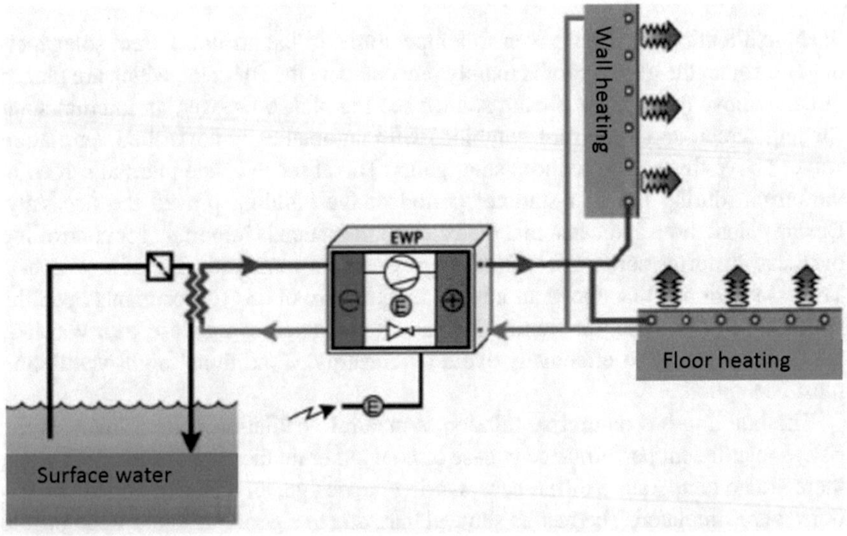

Fig. 4.7 Heat pump with surface water as source for heating and cooling

A heat pump on canal water as heat source delivers heat supply for the low temperature wall and floor heating system. The program of requirements described for the thermal comfort a maximum of 120 h above 25 °C and a maximum of 20 h above 28 °C, based on a reference climate year. The goal was to avoid a mechanical cooling system. The internal heat load is 33 W/m^2. Calculations indicated that with adequate solar shading, sufficient internal mass, and a good operating ventilation system, the required targets could be reached. The addition of internal mass to the wooden building was necessary to smooth peak temperatures during summer period. An additional 30 mm layer of loam plaster to the walls and the ceiling brought the solution. For the heat generation, originally, the heat pump releases heat through an open heat exchanger to the water of the canal; see Fig. 4.7.

This open heat exchanger caused a lot of problems to corrosion problems caused by the brackish water in the canal. As a result, the efficiency was low; now it has been replaced by a closed loop system which is filled with glycol and has a much better efficiency. The heat pump provides warm water for the heating system of 50 °C. Important for the efficiency is that the water of the canal is not too cold. If there is a hard winter, the water in the canal can reach nearly the freezing point, which is not a good thing for the efficiency of the system. Caused by the economic crisis, the industry connected to the canal had put less waste heat in the canal. Therefore, it is necessary to heat up the water before being supplied to het heat pump. Also, selection of heat exchanger and filter installation is important to reduce the effects of corrosion caused by the brackish water.

4.1.2 Materials

The main building structure is made out of wood and many other materials are recycled and renewables such as loam stone, cellulose insulation from old newspapers, natural paints, etc. Many of the building's materials are reused waste materials, for example, old mooring posts are used for facade cladding, stairs, etc. The wood used for the interior walls and the frames in the exterior wall have been treated using natural linseed oil-based paint. The carpeting is made from wool and goat hair, and the desktops are made from bamboo treated with linseed oil and beeswax.

4.1.3 Water Concept

The design seeks to buffer rainwater and to minimize the use of drinking water and wastewater produced. All wastewater in the Rijkswaterstaat offices from taps and showers to dishwashers and toilets is led through a collecting tank, where the first breakdown processes and homogenization occur, and into the reeds; see Fig. 4.8. The treated wastewater is used both for flushing toilets and for the water art. Water is used in the atrium: the gentle babbling of flowing water helps to create a pleasant atmosphere as well as contribute to adiabatic cooling. All the wastewater is treated locally, in two beds of reeds, and recycled for flushing toilets. The office was given a separate treatment system in the form of a helophyte filter. Much of the roof water is buffered by the moss-sedum roof, while the surplus is allowed to infiltrate through a soak way pit. Paving has been kept to a minimum around the building. Where paving is unavoidable, various forms of porous paving material have been used, such as gravel and woodchips.

The system's environmental yield is high, since it does not produce any wastewater, and the amount of drinking water consumed is cut by two-thirds. The absence of

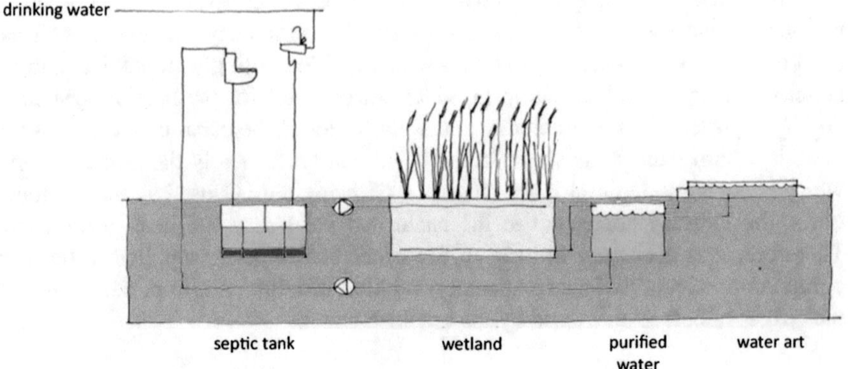

Fig. 4.8 Scheme of the water system © atelier GROENBLAUW

pavement and the addition of a sedum roof offer benefits in both climatological and ecological terms.

Architect: opMAAT, Delft
Consultant energy concept: Cauberg-Huygen, Amsterdam
Consultant building services: Bravenboer en Scheers, Terneuzen
Consultant structures: Bliek en Vos, Terneuzen

4.2 Biesbosch Museum

In November 1421 the St. Elizabeth's flood caused the loss of De Grote Waard in the southwest of the Netherlands. The dikes broke in a number of places and the lower lying polder land was flooded. Villages were swallowed by the flood and were lost, causing between 2000 and 10,000 casualties. It created the Biesbosch, now a valued nature reserve as part of a national water safety program; the 4450-ha Noordwaard polder has been turned into a water retention area. The history of the importance of water safety was central at the development of the Biesbosch Museum, and the permanent exhibition explains the historical development of the region from the Elizabeth Flood of 1421 to its current status as a recreational area. The museum is surrounded by earthworks and covered with a roof of grass and herbs; see Figs. 4.9 and 4.10. It is designed to minimize energy consumption. The roof adds ecological value, creating a sculptural object that reads as land art and, at the same time, manifests itself in the surrounding landscape. The earthworks on the northwestern side and the green roof serve as additional insulation and a heat buffer.

Fig. 4.9 Side view of Biesbosch Museum

Fig. 4.10 Back side of Biesbosch Museum, transparent and open to the activated educational garden

4.2.1 Energy Concept

The heating is done by two 55 kW wood boilers using wood pellets as fuel. The heating installation has a buffer tank of 2000 L to supply floor heating, convectors, heating panels, and the heating coil of the air handling unit. The glass front is fitted with state-of-the-art heat-resistant glass. On cold days, a biomass stove maintains the building at the right temperature through floor heating. On warm days, water from the river is used to cool the building by means of an electrical heat pump; see Fig. 4.11. The cooling system has a storage tank of 1000 L for the floor cooling, convectors, and cooling coil of the air handling unit. Normally, the temperature trajectory is 18–26 °C during the summer; the heat pump changes this to 6–12 °C to use the river cooling as much as possible.

4.2.2 Materials

To avoid any unnecessary waste of material, the structure of the original Biesbosch Museum has been retained, and a new 1000 m^2 wing was added on the southwestern side of the building. The museum uses many available resources in the area, natural materials from the Biesbosch itself, as much as possible.

4.2.3 Water Concept

Sanitary wastewater is purified through a willow filter: the first in the Netherlands and an acknowledgment of the wicker culture of the Biesbosch. Willows absorb the wastewater and the substances it contains, among them nitrogen and phosphate. These substances act as nutrients and help the willow to grow. The purified water is discharged into the adjacent wetland area and flows from there into the river.

Architecture: Studio Marco Vermeulen, Rotterdam
Building management: Edion, Driebergen
Consultant installations: Overdevest Adviseurs, Den Haag
Consultant building physics and fire prevention: MoBius Consult, Driebergen
Consultant structure: W5A Structures, Waalre

4.3 Restaurant Aan Zee, Oostvoorne

The building "Aan Zee" was erected on a dike about 9 m above the water level and was completed in 2011. It is situated in a sand dune conservation area in Oostvoorne, a protected flora and fauna area, making the building a unique symbol of sustainability. It is a new type of restaurant as it is nearly autarkic. The design by Emma Architecten is strongly inspired by the surrounding landscape, from which the shapes and colors are taken and interpreted. The design project was truly integral as it included landscaping, building, installations, interior design and design of food, menu, styling, graphics, and even clothing. After a design period from 2006 to 2010,

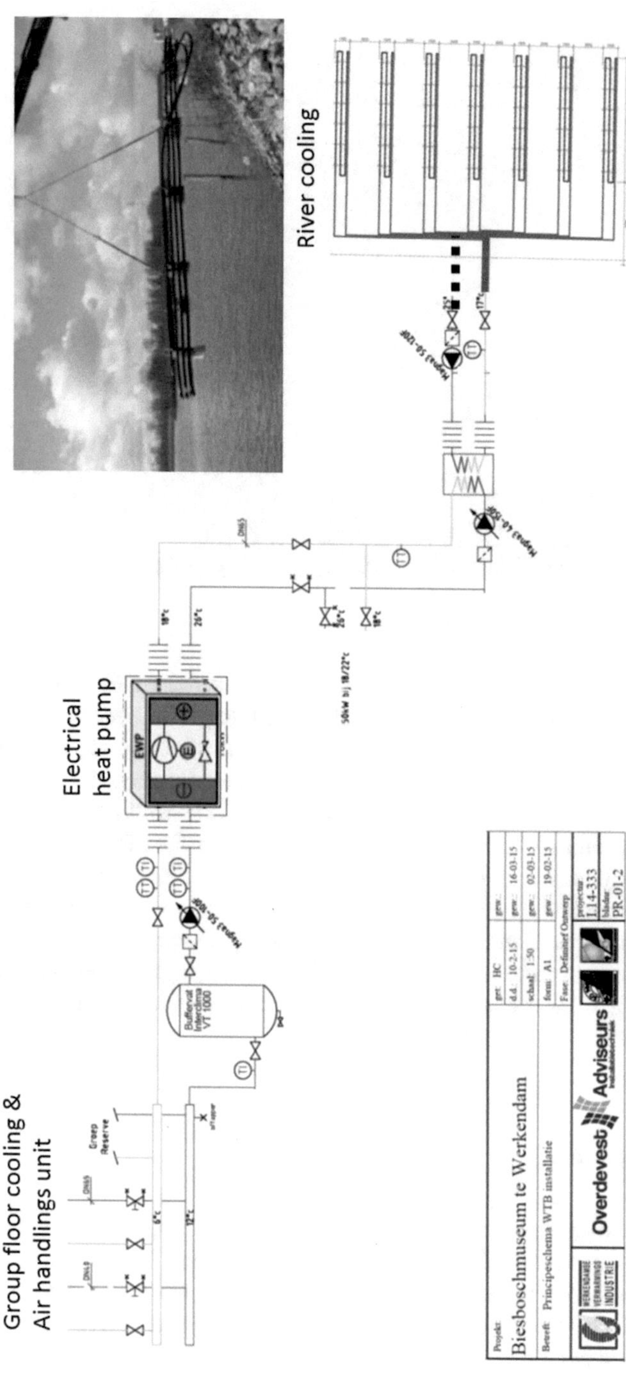

Fig. 4.11 Energy concept of Biesbosch Museum

Fig. 4.12 Front view of Restaurant Aan Zee

the restaurant was built within 4 months and was opened in 2011. Because of its unique concept, it was difficult for the entrepreneur to get the necessary permits that was the reason why the planning phase took so long. The help of the municipality was essential for getting the permits. The project achieved several recognitions and awards, such as the QualityCoast Gold Award (2013) as well as several awards in the hospitality industry for its unique design and concept. The Restaurant "Aan Zee" is sustainable in every respect – from the design and integration in the landscape to the constructive and technical structures of the energy and water utilization and all the way to the preparation of the meals. The building is impressive as well as expressive; its sculptural form fits discreetly into the surroundings, makes sustainable use of resources, and functions almost in perfect harmony with the environment in the long term; see Figs. 4.12 and 4.13.

4.3.1 Materials

The main building structure is made out of wood and many other materials are recycled and renewables. The back portion of the building is anchored in the dunes with old shipping containers, while the front half is a curving form with a wall of glass to take in the sights. A slatted wood rain screen tops the building to help shield the sun and weather, further aging over time to blend with the landscape. This building was constructed with natural materials and an innovative design in order to

Fig. 4.13 Side view Restaurant Aan Zee

ensure its harmonious integration into the dune landscape. Built in only 4 months, it can be easily dismantled and recycled. It is reusable both as a building and on a level of materials.

4.3.2 Energy Concept

The restaurant has been designed to be as energy efficient and sustainable as possible through the use of efficient lighting and appliances, solar and wind energy, natural supply ventilation, and even geothermal techniques for heating and cooling. Solar panels have been installed in the roof and visitors can climb to the top of its spiraling watchtower to enjoy the views. Electricity is generated by two windmills, each with 5 kWp, and photovoltaic elements on the roof, which produce about 13 kWp. To save energy, only LED lighting is used throughout the building. The heat is generated by a geothermal heat pump with a maximum coefficient of performance (COP) of 4.4. Eleven probes at a depth of 95 m provide a total of 2090 m of vertical storage; see Fig. 4.14. In combination with eight solar collectors (total 18.8 m^2, 13 kWp) and an 1500 L storage system, the natural heating and cooling of the building is ensured, as is the hot water supply by means of solar and geothermal

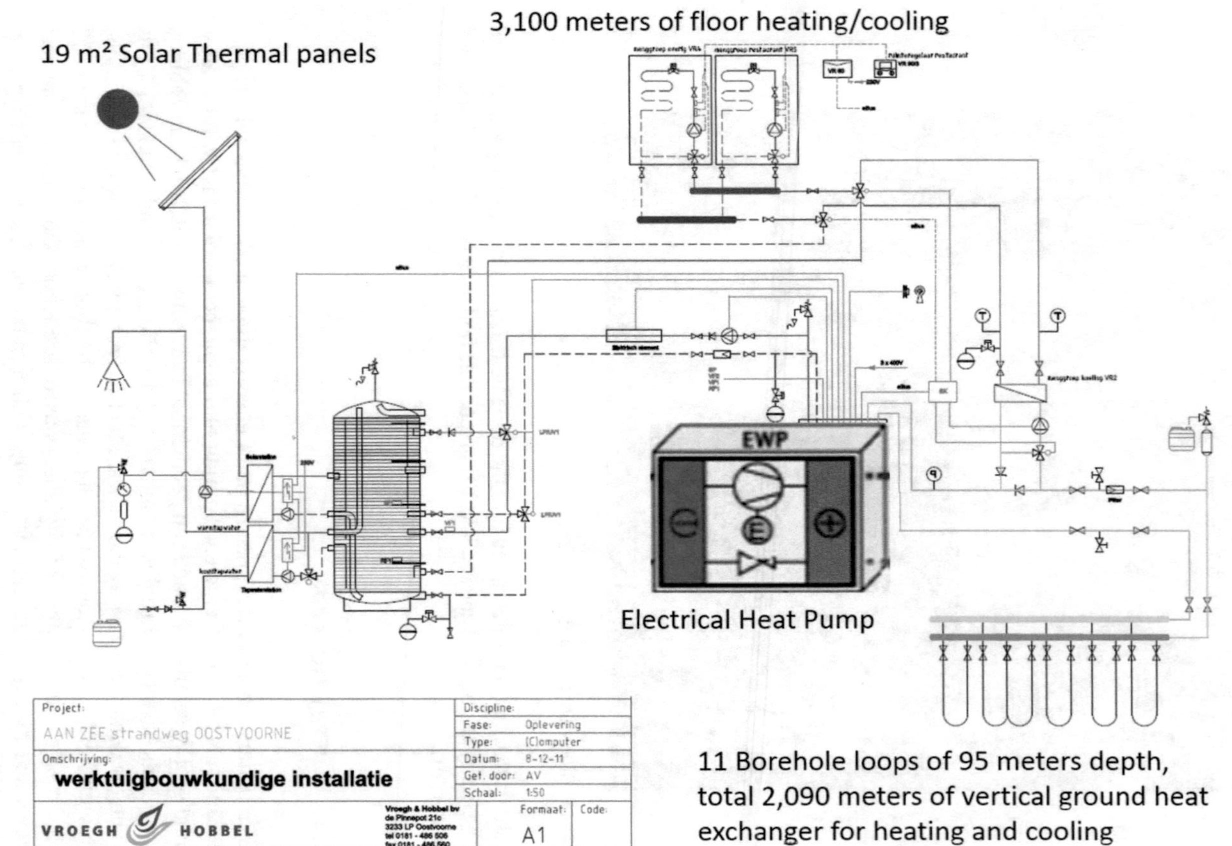

Fig. 4.14 Energy concept for heating and cooling of Restaurant Aan Zee

heat. The distribution of heat and cold in the building itself is achieved by an in-floor heating/cooling system. For this purpose, 3.100 m of in-floor heating were installed throughout the building, also in the containers. The open fireplace primarily serves to create a cozy atmosphere but, in winter, also helps to preheat the air supplied to the building. Ventilation of the building is achieved with a natural convection system.

4.3.3 Water Concept

A special feature of the building service engineering concept is the water management. The wastewater is cleaned naturally in a 200 m^2 halophyte field, which eliminates the need for connection to the sewage system. First, the fatty components are filtered from the wastewater and are later used to supply a biomass heating system (Fig. 4.15).

Architect: Emma Architecten, Amsterdam
Construction management: Jan Lemmens BV, Oostvoorne
Construction: SvR bouwconstructies, Spijkenisse
Contractor: I. van Reek en Zn Bouw en Aannemingsbedrijf Oostvoorne

4.4 Energy from the Sea Duindorp Scheveningen

A complete different approach to architecture and the sea was taken in this project. A small nondescript white building housing the power plant sits just inside the industrial Duindorp's harbor complex; see Fig. 4.16. However, the innovative approach to sustainability for its energy-saving concept stands out as a promising solution for waterfront communities in the clean energy transition (Marcaccii 2014; Foster 2014). When thinking of harnessing renewable energy from the ocean offshore, wind farms are the first thing that comes to mind. However, the oceans as a vast source of renewable heat can be used to keep homes warm. Remember, more than 70% of the Earth is covered with water and serves as a kind of global thermostat. Harnessing just a tiny fraction of the heat stored in the world's oceans does not seem a bad idea. In Duindorp, a small harbor town near The Hague in the Netherlands the Dutch environmental engineering firm Deerns International designed an innovative and sustainable district heating system fed by the heat of the sea. District heating using seawater promised to be the most affordable solution, insuring no resident would have to pay more than the national average of €70 (about $94) a month for heat and hot water. A small warehouse-looking building contains both the central heat exchanger and heat pump. The district heating system warms water at a central location and then distributes it through a system of underground pipes.

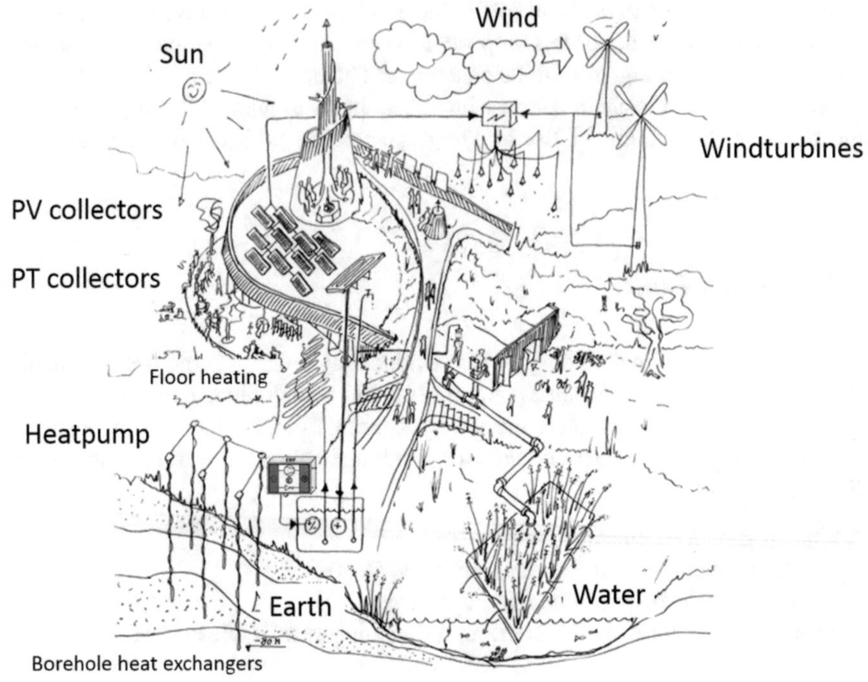

Fig. 4.15 Total sustainability concept of Restaurant Aan Zee [Emma Architecten]

4.4.1 Energy Concept

None of the water in the pipes is used directly in homes, but the heat from the water is skimmed off and used by individual heat pumps in the houses to warm showers (maximum 60 °C) and floors (maximum 45 °C); see Fig. 4.17. The individual heat pumps are installed in each home for further heating and due to the energy concepts are very small and easy to install within the buildings (Table 4.1).

In the summer, creating warm water to flow through the district heating network of pipes is relatively straightforward. Intake pipes at the harbor draw in about 95,000–190,000 L of warm seawater every hour. An extensive series of filters throughout the intake system ensures that no sea life is sucked into the district heating plant. The seawater is used in a heat exchanger to heat freshwater for the pipes to around 12 °C, resulting in 11 °C in the distribution net. The warmed freshwater is then sent out along a five-mile network of insulated pipes that services the 789 homes in the new housing neighborhood. At every house connected to the system, a 5-kWh-capacity heat pump raises the temperature of the water between 40 and 65 °C for heating and warm water. Annually, this warm water accounts for almost one-third of the total heat demand. The houses are cooled as well: the floor heating system is used during summer to cool the houses. The heat that is retrieved from the

Fig. 4.16 Central heat pump using energy from the seawater to distribute it by underground pipes toward the houses

houses is fed into the distribution network. On hot days, in turn, the network is cooled by the sea.

The source for the heat pump in the houses is a distribution piping network; see Fig. 4.17. This distribution network is kept on a temperature of 11 °C. This is done by the seawater heating plant; see Fig. 4.18. In the distribution network, water circulates toward the houses and back to the plant. The SWHP heats the distribution network, in summer with a high seawater temperature directly by a heat exchanger and in winter by the heat pump. Seawater is being pumped up from the harbor. For most of the winter, the temperature in the harbor is right around 2–5 °C and in summer it can climb to near 21 °C. 0 °C has appeared only once: in the extreme winter of 1963.The seawater is pumped through the *inox pipes* toward the plant building where it is filtered further before it is going to be used as a heat source. The pumps have a capacity of 250 m^3/h each. The flow normally varies from 80 to 500 m^3/h. At extreme low seawater temperature condition (0 °C or lower), the flow can be boosted to 750 m^3/h.

In the winter, the operation of the system is more complex to reach energy efficiency. Unfortunately, the moment when you need heating the most, the ocean is at its coldest, sometimes just around 2 °C. That immediately affects the performance of the heat pump. Much more water is needed in the winter, compared in the summer, to keep the system running. About 190,000 gallons of water is taken in every hour when the ocean is at its coldest and only a few degrees of heat can be transferred. If the water is just a few degrees above freezing, much more flow is needed.

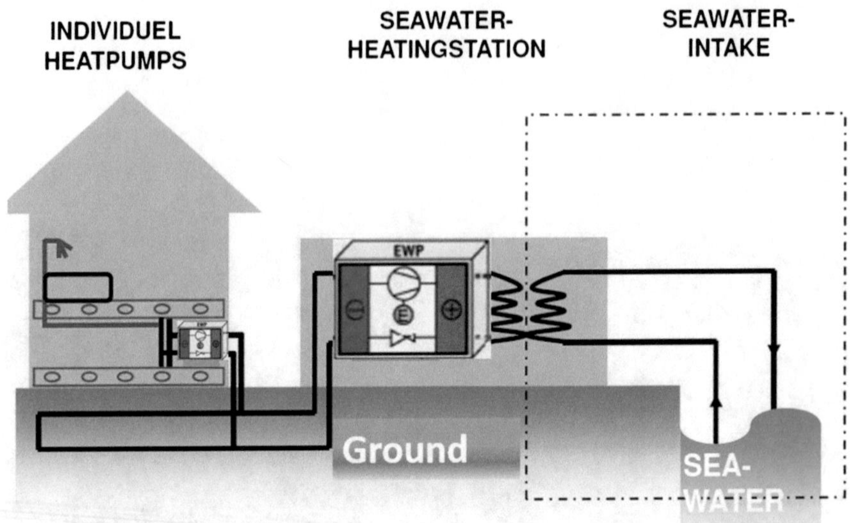

Fig. 4.17 Concept of individual heat pumps in de houses connected to the central heat pump using energy from seawater

At this stage, the central heat pump has to deliver full power; it starts already when the seawater is below 11 °C. The electricity needed to run the heat pumps is about 3 MW. The system is extremely efficient generating 15 kWh of heat for every 1 kWh of electricity pumped into the system. This reduces carbon emissions by 50% when compared to conventional heating using natural gas. While district heating systems and heat pumps certainly aren't new ideas, making the system run smoothly on the energy from the sea with an affordable price tag was a massive undertaking. Dealing with huge volumes of very salty water in the mechanical systems every day is problematic. Finding ways to battling the problem of corrosive seawater were necessary to cut down on costs of replacing corroded components. Seawater district heating will be most cost-effective in areas where new development is taking place. Since district heating depends on an underground network of pipes, retrofitting a community to run on a district heating system would add considerably to the price tag. This design would work especially well and cost less if the community was near a large body of freshwater, when you don't have to worry about saltwater ruining the equipment. Any town or city on the coasts, along the Great Lakes, or even near large rivers could benefit from a similar system.

Project management: CERES projecten
Consultant energy concept: Deerns Consulting Engineers BV

Table 4.1 Concept of energy (Foster et al. 2016)

		Central heat pump		Individual heat pumps
		Winter	*Summer*	*Year-round*
Installed plant	Number of heat pumps	Two central heat pumps		One per apartment
	Refrigerant	Ammonia		Not known
	Heat pumps used	York PAC 163HR		IVT (no longer in business)
Heating	Heat supply temperature	11 °C	HP not used when sea above 11 °C	Underfloor at 45 °C DHW up to 60 °C
	Source	Seawater	–	District heating network
	Source temperature	3 °C	Up to 20 °C	11–20 °C
	Max heat output	2.4 MW		6 kW per apartment
Cooling	Cooling provided?	No		No
Performance	COP (heating)	11	Heat pump not used	3

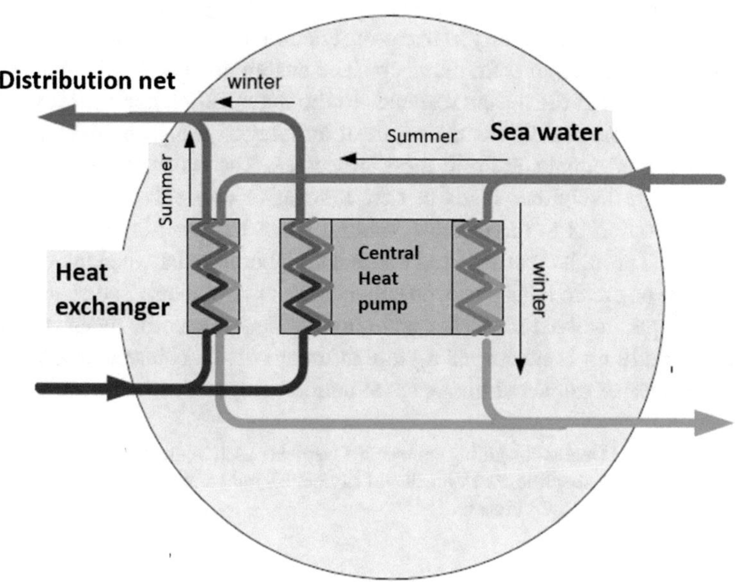

Fig. 4.18 Concept of the seawater heating plant and its central heat pump

4.5 Conclusions

The Netherlands is a small but highly populated country next to the sea behind dunes and with some relatively big rivers. Therefore, it has a special relationship with water. Careful building service design can minimize the need for heating and cooling throughout the year, for example, by applying nearby rivers and the sea as a source of heat pumps. One of the first projects was the office building of the governmental sea and river control agency RWS at Terneuzen. Another interesting building is the Biesbosch Museum were a part of the Dutch history of the fight against water can be seen and where now the water is used as an energy source for the heating and cooling. Furthermore, some interesting examples are presented by applying the surface water of the North Sea in projects in The Hague and Oostvoorne. It all shows the Dutch innovative ways of not only fighting against the water but also using water as a sustainable source of energy.

Sustainability is more than just saving energy and concerns a set of aspects: surrounding, connectivity, material use, wastewater, health and safety, energy flows, and last but not least beauty. The balance of all these aspects determines the degree of sustainability of a building. Moreover, if one of these aspects is insufficiently considered, the building will not be sustainable and might be demolished prematurely. Therefore, the design process must be a broad exploration on all aspects and in the dialogue with principle, and users should be clarified what needs to be considered most. In all projects presented here, there was always an intensive dialogue with the client to realize the ambition to build.

The Dutch building design process is characterized by creating widespread support for the project, particularly in terms of function, through extensive consultation with users and the design team members. The design team with all the consultants involved ensures that the design responds to the organization and how it works. An optimal sustainable building is the result of an integral design process. All design disciplines are of equal stature in these processes. The synergy between different building design disciplines leads to new innovative concepts and solutions. The main consideration is to select those solutions that have minimum impact on the environment. Through creativity and optimal collaboration between the disciplines, solutions can be created that can even enhance the environment. The changed mindset of all people involved in the integral building design process means that a truly sustainable building is no longer a utopian dream. Architecture and the sea offer some innovative examples of the Dutch situation.

Acknowledgments Deerns consulting engineers, Overdevest Adviseurs, I. van Reek en Zn Bouw, Aannemingsbedrijf Oostvoorne, and Vroegh and Hobbel BV, all contributed to this article by providing information about their projects.

References

de Flander, K. (2011). *Office building RWS Terneuzen*. http://resourceculture.de/articles/office-building-rws-terneuzen

Foster, J. M. (2014, July 24). This town is using the ocean to provide heat to low-income residents. *ThinkProgress*. https://thinkprogress.org/this-town-is-using-the-ocean-to-provide-heat-to-low-income-residents-94a608e8a177#.4f2ycwysx

Marcaccii, S. (2014, June 24). *Sustainable district heating from seawater? It's happening in Holland.* https://cleantechnica.com/2014/06/24/sustainable-district-heating-from-seawater-its-happening-in-holland/

Pötz, H., & Bleuzé, P. (2009). *Vorm geven aan stedelijk water.* Amsterdam: SUN Architecture. http://www.urbangreenbluegrids.com/projects/rijkswaterstaat-office-in-terneuzen-the-netherlands/

van der Aa, A., Eijdems, H. H. E. W., & Cauberg, J. M. M. (2002). Office building Ministry of Transport at Terneuzen, The Netherlands: Low energy, high thermal comfort and high sustainability in one concept. In *Proceedings of Indoor Air 2002 (9th International Conference on Indoor Air Quality and Climate)*, June 30–July 5, 2002, Monterey.

Chapter 5
Sustainability Measures of Public Buildings in Seaside Cities: The New Library of Alexandria (New Bibliotheca Alexandrina), Egypt

Mohsen M. Aboulnaga

Image credit: www.bibalex.org

M.M. Aboulnaga (✉)
Department of Architectural Engineering, Faculty of Engineering, Cairo University, 12613 Giza, Egypt
e-mail: mohsen_aboulnaga@yahoo.com

A. Sayigh (ed.), *Seaside Building Design: Principles and Practice*, Innovative Renewable Energy, https://doi.org/10.1007/978-3-319-67949-5_5

Image credit: Author

Objectives

By studying this chapter, you should be able to:

- Understand some typical features of seaside or waterfront libraries.
- Comprehend various approaches to develop and design public libraries.
- Gain knowledge about design principles of seaside libraries.
- Recognize the main elements in making libraries sustainable.
- Analyse the sustainability principles and design elements applied in the Bibliotheca Alexandrina.

> Cities embody some of society's most pressing challenges, from pollution and disease to unemployment and lack of adequate shelter. But cities are also venues where rapid, dramatic change is not just possible, but expected. (Ban Ki-moon, the Secretary-General of the United Nations)

5.1 Introduction

Seaside cities worldwide are facing enormous challenges. One of the major challenges is how these cities adapt to severe events caused by climate change impacts. According to the IPCC climate model prediction, many cities, such as Miami (Florida), Brighton (UK), Venice (Italy), Barcelona (Spain), and Istanbul (Turkey), as well as Tunis (Tunisia), Alexandria and Port Saeed (Egypt), and Dubai (United Arab Emirates), are vulnerable to climate change risks.

Coastal cities exist in different parts of the world. According to the classification of the best 25 coastal cities in the world in reference to the living conditions and

North America Canada & USA	Europe	Asia	Australasia Australia & New Zealand
New Orleans.	Edinburgh	Bangkok	Sydney
Honolulu	Venice	Dubai	Perth
Miami	Istanbul	Hong Kong	Auckland
Los-Angeles	Oslo	Tokyo	
New-York City	Lisbon	Singapore	
San Francisco	Helsinki		
Chicago	Dublin		
Vancouver	Copenhagen		
	Barcelona		

Fig. 5.1 List of well-known coastal cities worldwide (Chart's credit: Author)

quality of life. Figure 5.1 presents these cities in different continents. This ranking was carried out using tools and analysis to classify such cities based on these criteria.[1]

> Seaside or Waterfront Public Libraries need to address sustainability measures more than non-seaside libraries, due to severe events caused by climate change risks. (Author)

Image source: https://commons.wikimedia.org/wiki/File:Danish_Royal_Library.jpg

[1]List of the top 25 coastal cities in the world. Available at http://www.cntraveler.com/galleries/2016-06-24/the-25-best-coastal-cities-in-the-world. Accessed on March 4, 2017

Box 1: Seaside Libraries Worldwide

- The John F. Kennedy Presidential Library and Museum, Boston, Massachusetts, USA
- The Library at the Dock, Melbourne, Australia
- The Royal Danish Library (the Black Diamond), Copenhagen, Denmark

5.2 Libraries in Seaside Cities Worldwide

There are many libraries worldwide that are overlooking seaside or waterfront areas (rivers). The libraries which are to be assessed and discussed are in the United States, Denmark (Europe) and Australia. Each of these libraries is located in a different climatic zone. Figure 5.2 shows images of these libraries. It has been noticed that the sustainability measures in these three libraries vary according to each library's date of construction. Table 5.1 sets the criteria for the comparison between the three waterfront libraries and highlights the differences between them.

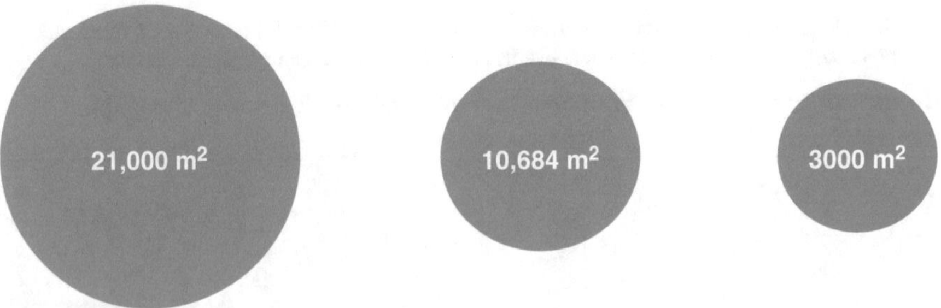

The following section describes each of the three waterfront libraries in terms of four features:

(i) Concept and form
(ii) Spatial experience
(iii) Building conditions
(iv) Sustainability

5.2.1 *John F. Kennedy Presidential Library and Museum, Boston, USA*

The first of the three waterfront libraries is the John F. Kennedy Presidential Library and Museum in Boston, Massachusetts, USA. The library represents a memorial of the 30th President John F. Kennedy and is overlooking the sea – a waterfront – and

a.

b.

c.

Fig. 5.2 Waterfront libraries in different continents (United States, Europe and Australia). (**a**) John F. Kennedy Presidential Library (Image source: https://www.jfklibrary.org/Asset-Viewer/UtECQ0uHGUSP6dWsB4MGow.aspx). (**b**) The Library at the Dock (Image source: http://www.architectureanddesign.com.au/news/australia-s-first-cross-laminated-timber-clt-publi). (**c**) The Royal Danish Library (Image source: https://bluesyemre.com/2015/08/19/27-libraries-to-visit-in-your-lifetime-by-sarah-schmalbruch/also-known-as-the-black-diamond-because-of-its-black-granite-and-unusual-angles-the-royal-library-of-copenhagen-is-an-extension-of-the-national-library-of-denmark-the-librarys-glass/)

is located on a ten-acre park. This Library is a significant work of late modern architecture which was designed by I.M. Pei & Partners of New York. The site in Columbia Point was selected in 1976 and the project was completed in 1979. Upon completion, the library consisted of ten levels. Later in the early 1990s, a west expansion was added and some renovations were made.[2] The monumental building stands at the end of the peninsula in subtle yet effective way representing a waterfront landmark, setting an example for an architectural memorial[3] (Fig. 5.3).

[2]About the John F. Kennedy Presidential Library. Available at https://www.jfklibrary.org/About-Us/About-the-JFK-Library.aspx. Accessed on March 14, 2017

[3]Greg LeMaire. "AD Classics: AD Classics: JFK Presidential Library/I.M. Pei" August 07, 2011. ArchDaily. Available at http://www.archdaily.com/153285/ad-classics-jfk-presidential-library-i-m-pei. Accessed on March 14, 2017

Table 5.1 Three waterfront libraries in the United States, Europe and Australia

Item	John F. Kennedy Presidential Library	The Royal Danish Library	The Library at the Dock

	Image source: https://www.jfklibrary.org/About-Us.aspx	Image source: https://commons.wikimedia.org/wiki/File:Danish_Royal_Library.jpg	Image source: https://sourceable.net/library-at-the-dock-australias-first-public-clt-building-opens/
City	Boston, Massachusetts, USA	Copenhagen, Denmark	Docklands, Melbourne, Australia
Year	1979	1999	2014
Architect	I.M. Pei	SHL Architects	Clare Design
Area	21,000 m^2	10,684 m^2	3000 m^2
Height	10 floors, 38 m (tower)	7 floors, 29 m	3 floors, 12 m
Description	A memorial library for the United States' 35th President, John F. Kennedy	A modern library linked to Hans J. Holm's old library cathedral, 1906	A sustainable modern library that acts as a community hub for the area at Victoria Harbor
Building style	Post-modern	Post-modern	Contemporary
Climate	Humid continental climate, with warm summers and cold snowy winters	Oceanic climate, with rainy summer and snowy winter	Temperate oceanic climate (well known for its changeable weather conditions)
Form	A single protruding triangular volume from a base tower	A complex yet subtle form that contrasts with the old Holm Library	A timber cladded box with a large awning projecting outwards
Masses	Geometric masses	A strong solid black mass with a large area of glazing dividing the building into two parts	Lightweight and simple yet beautiful building in a form of single box
Cost *(EUR)*	€ 11.25 million	€ 49 million	€ 9.3 million
Sustainability principles	None	None except for natural daylighting and social sustainability	Economically, socially and environmentally friendly. Constructed with cost-effective environmentally friendly materials. Solar PV and rainwater treatment. The library also acts as a social hub

a. *b.*

Fig. 5.3 John F. Kennedy Presidential Library. (**a**) The library on the waterfront (Image source: http://www.architravel.com/architravel/building/john-f-kennedy-presidential-library-and-museum/). (**b**) The shape of the library's triangular form (Image source: https://www.jfklibrary.org/About-Us.aspx)

5.2.1.1 Concept and Form

The JFK Presidential Library has a simple and unique yet strong form. The monumental scale of the building is an excellent way of reviving the memory of the late president which is vividly shown in the building form. The main element of the building is the singular protruding triangular tower, which is contiguous to a glass and steel cube and stands on a base of geometric forms. The steel and glass form of the building allows for connection with the outside environment and links the library with the state which President Kennedy dedicated his life to.[3]

5.2.1.2 Spatial Experience

The library design in terms of users' circulation makes the visitor experience the memory of the president. Following this, the viewers find themselves entering a theatre showing a brief film about the president.[3] The nine-storey tower is 125 feet high and consists of educational spaces, archives and administrative spaces. The two-storey base accommodates for an 18,000 square foot exhibition area, 230-seat theatres and a glass memorial pavilion of a 115 feet height. The sun-filled glass and steel pavilion of the façade gives the visitors a different experience after entering from a less-lit zone. The space is designed to allow for a panoramic view of the sky, land and sea through the glass and steel structure.[4] Nonetheless, a 21,800 square foot extension was added to the Kennedy Library in 1991 – known as "Stephen

[4]About the John F. Kennedy Presidential Library, I.M. Pei, Architect. Available at https://www.jfklibrary.org/About-Us/About-the-JFK-Library/History/IM-Pei--Architect.aspx. Accessed on March 14, 2017

E. Smith Centre". The two-storey annex was designed by the same architect, I. M. Pei., in the same style as the original library. It accommodates lectures, conferences, meetings and archival functions.[4]

5.2.1.3 Building Conditions

There were some constraints facing the construction of the library, mainly the current location. First of all, the library would be built on a waterfront site which was originally a landfill. To solve this issue, the site was raised 15 feet. Secondly, the design had to meet a programme that satisfies a mixed-use building and fulfil the aim of creating a contemporary memorial to the late president.[4]

Box 2: Seaside Libraries' Design
The architect has successfully dealt with this condition by creating a split-level design where the museum spaces lie in an underground level.[4]

5.2.1.4 Sustainability

Sustainability measures were not taken into account while designing and constructing this library which is normal, given that sustainability was not as important in the time when the library was built, which was 15 years ago and is, however, of a prime importance now.

5.2.2 The Royal Danish Library "the Black Diamond", Copenhagen, Denmark

Another example of waterfront libraries is The Royal Danish Library, also known as the "Black Diamond". The library is a modern library in the city of Copenhagen. The waterfront building was designed by Schmidt, Hammer and Lassen, the Danish architects. It lies adjacent to but contrasts in the form and materials – the 1906 Hans J. Holm cathedral library.[5] The "Black Diamond" name refers to the library's prismatic form and the black marble material which is integrated with the glazed facades. The Royal Library is considered a cultural centre and a part of the harbour promenade.[5] The new extension allowed for increasing the library's capacity, which reaches 200,000 books now.[6]

[5] Black Diamond, Royal Danish Library, Librarybuildings.Info. Available at http://www.librarybuildings.info/denmark/black-diamond-royal-danish-library. Accessed on March 14, 2017

[6] The Royal Library, Schmidt Hammer Lassen Architects. Available at http://www.shl.dk/the-royal-library/. Accessed on March 14, 2017

Fig. 5.4 The Royal Danish Library (Image source: http://www.arcspace.com/features/schmidt-hammer-lassen-architects/the-black-diamond/)

5.2.2.1 Concept and Form

The library consists of a black prism with a huge glass feature dividing the building into two parts. It contrasts the old Holm Library in the form and materials.[7] The building's form is leaning towards the river. To achieve stability of the library's complex form, the building's angles were accurately calculated. The black glazing allows daylighting to enter into the entire building and provides a panoramic view of the waterfront. Also, the glazing of the ground floor gives the impression that the building is floating and allows for a perfect view of the waterfront (Fig. 5.4).

The old library (the Holm building) is connected with the new library with the Christians Brygge thoroughfare.[8]

5.2.2.2 Spatial Experience

The building, which consists of seven stories – excluding the basement, provides an excellent illustration of good spatial experience, starting from the balconies in the central foyer that provide a panoramic view of the harbour to the connection with the old building. The foyer also acts as a central cultural hub for the public. The library includes reading rooms, a concert hall, exhibitions, and a bookshop as well as a café and restaurant.[5] The interior design of the atrium contrasts with the exterior form. It is designed in an organic way with wave-like balconies and a 24 m height. In addition, its massive skylight provides indirect day lighting for the reading rooms and allows views not only from the inside but also from the outside to the library's

[7] The Black Diamond, Arcspace. Available at http://www.arcspace.com/features/schmidt-hammer-lassen-architects/the-black-diamond/. Accessed on March 14, 2017

[8] Monolithic Modern Design-The Royal Library In Denmark By Shl Architects. Homesthetics.net. Available at http://homesthetics.net/the-royal-library-in-denmark-by-shl-architects/. Accessed on March 14, 2017

beautiful interior. As shown in Fig. 5.5, the atrium is directly connected to different spaces within the library, including the bookshop, the cafe/restaurant, the Queens Hall, and the exhibition galleries.[7] The old library's entrance is linked with the Black Diamond's atrium by a glass passage as illustrated in Fig. 5.6a.[7] There had to be a transition in the link between the two libraries since these are different architectural

Fig. 5.5 The Royal Danish Library – atrium and galleries (Image source: http://www.shl.dk/the-royal-library/)

a.

b.

Fig. 5.6 The Royal Danish Library. (**a**) The link between the two libraries from the outside (Image source: https://danish.tm/article/architecture-lasts-black-diamond/). (**b**) The link between the two libraries from the inside (Image source: http://www.shl.dk/the-royal-library/)

Fig. 5.7 The Royal Danish Library – travellators (Image source: https://commons.wikimedia.org/wiki/File:Black_Diamond_-_travelators.jpg)

styles (post-modern and classic). To achieve this, the artist, Per Kirkeby, enhanced the link between the two buildings with a painting and a large circular canvas[6] as shown in Fig. 5.6b.

The building's walking distances are kept short using a mean of mobility for visitors and staff. This is in the form of two travellators that connect between the entrance level and the third floor. These travellators are linking the main building's floor with the old library to reach the 18-meter-wide bridge above Christians Brygge[7] (Fig. 5.7).

5.2.2.3 Building Conditions

While the old library is made of a brick structure (classic style), the new library is built using steel framing (post-modern) showing a clear contrast between old and new.

> **Box 3: Linking the Old Library to the New Library**
> The lesson learned from this example is how to link old buildings to new ones.

5.2.2.4 Sustainability

By examining the environmental sustainability of the library, it shows some features of sustainability such as the provision of natural daylighting and sound insulation. In terms of social sustainability, the library acts as a public focal point for the neighbourhood and the city. However, in terms of economic sustainability and financial sustainability, we don't have available data to judge on such pillars; thus we cannot at this stage confirm the sustainability of this library.

Box 4: Sustainability of Buildings
To judge on whether a building is sustainable or not, the following four sustainability pillars must be examined:

- Economic sustainability
- Social sustainability
- Environmental sustainability
- Financial sustainability

5.2.3 *The Library at the Dock, Melbourne, Australia*

The Library at the Dock demonstrates an exemplary model of waterfront libraries. The Library represents a new addition to Melbourne's Docklands precinct. It acts as both a library and a community centre by providing library and study functions as well as meeting, exhibition, and entertainment spaces.[9] Additionally, the Library at the Dock's design – awarded a 6-Star Green Star public building rating – respects sustainability features and is Australia's first building to be constructed using cross-laminated timber (CLT) technology.[9] This is the only one of the three libraries that is unique in terms of integrating sustainability principles.

5.2.3.1 Concept and Form

The library is designed in an elegant yet subtle building which stands on a heritage-listed wharf.[9] It is a three-storey timber clad box building with a canopy that provides shading for the whole front façade of the building and acts as a thoroughfare along the wharf.[10] The building entrance is transparent giving a welcoming feeling. The use of natural material connects the building with the surroundings, while its layered façade provides lighting and shading that changes the appearance of the library.[9]

Box 5: Sustainability of Buildings
This design demonstrates how to use natural materials that can give an excellent impression of buildings' design, especially in listed heritage sites.

[9] Library at the Dock, Hayball Projects. Available at http://www.hayball.com.au/projects/docklands-library/. Accessed on March 14, 2017

[10] The Library at the Dock, Arcspace. Available at http://www.arcspace.com/features/clare-design/the-library-at-the-dock/. Accessed on March 14, 2017

5.2.3.2 Spatial Experience

The library, which is used as a cultural hub for the community, has a simple and efficient design in terms of arranging its floors around the circulation core which provides daylighting through different levels within the library.[10] The ground floor, with a high ceiling, is overlooking a nearby park. It accommodates spaces for book display, entertainment and children's activities.[10] For the upper levels, the design is more horizontal with openings to allow for a panoramic view of the surrounding waterside, and the high windows bring natural daylighting into the inner spaces. These spaces are mixed use, including gaming areas, study areas and moveable shelves.[10]

Box 6: Sustainability of Buildings
The design of the library should show how the library spaces meet the required functions of a building.

5.2.3.3 Building Conditions

The building construction was a cost-effective one. Cross-laminated timber (CLT) was used in the upper floor slabs, the roof, as well as the walls. It is a beam and column construction system building. For the column and beam structure, GluLam (glued laminated timber) was used.[10]

5.2.3.4 Sustainability

The library represents an excellent example of socially, environmentally and economically responsible buildings. This is clear in respecting the idea of the users' needs and human scale. Its interior is designed in a way that allows occupants to feel familiarity and comfort with enjoyable reading and studying areas, and can also control their environment through the operable blinds and glass louvers.[10] The sustainability features included in the building are shown in Box 7.

Box 7: Sustainability of the Library

1. Use of cross-laminated timber (CLT)
2. Passive ventilation
3. Natural daylighting
4. Solar energy collection
5. Rainwater treatment

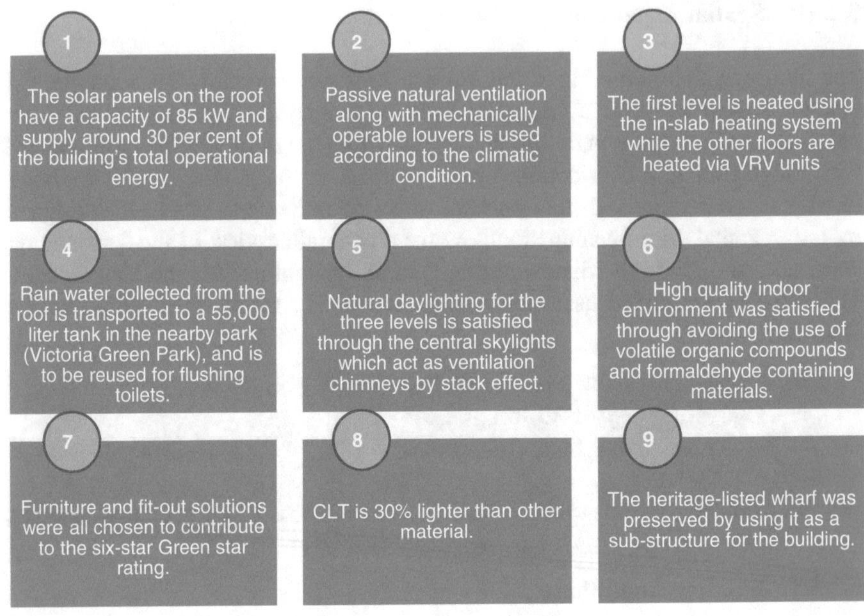

Fig. 5.8 Sustainability features in Library at the Dock, Australia

The use of CLT also contributed in reducing the buildings' weight which was in favour of the building location on the old wharf.[11] The sustainability features are detailed as follows (Fig. 5.8).[12]

It is clear from the three presented libraries that these buildings are classified as public buildings. The following section discusses the public buildings in terms of definition and type.

5.3 Public Buildings

Buildings are classified according to the type of function or activity the users perform, for example, residential, commercial, educational, medical, governmental, military, industrial or religious buildings. The building ownership can be considered as public or private. A public property does not belong to a specific person. Instead, it belongs to a city and is used by the public. A private property is owned by an individual, which gives the owner the right of use or sale of the property. Also,

[11] Cheng, L. (June 5, 2014). The Library at the Dock, Architectureau. Available at http://architectureau.com/articles/the-library-at-the-dock/#img=2. Accessed on March 14, 2017

[12] Library at the Dock, Claredesign. Available at http://claredesign.com.au/portfolio/public/docklands-library/docklands-library-info/. Accessed on March 14, 2017

Fig. 5.9 Classification of libraries (Image source: https://ebookfriendly.com/modern-libraries/)

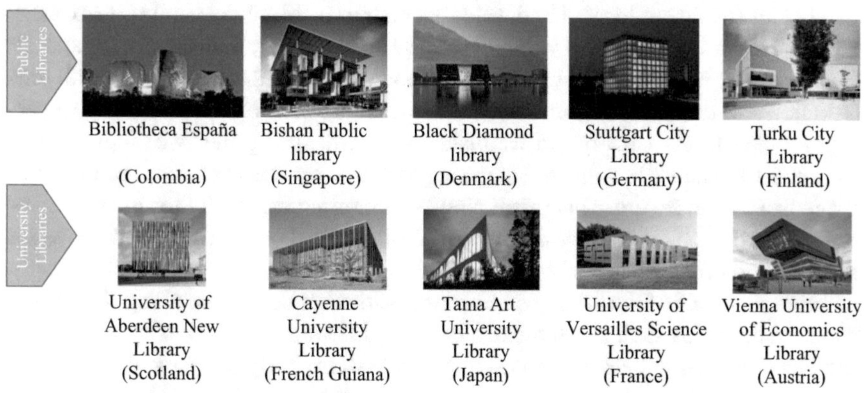

Fig. 5.10 Examples for public libraries (*top* pictures) and university libraries (*bottom* pictures) (Image source: https://ebookfriendly.com/modern-libraries/)

private properties' owners are subject to public rights such as paying taxes. Libraries are classified as educational buildings and can either be public or owned by an institution (school or university) as shown in Fig. 5.9. Public libraries are accessible by the general public such as Bibliotheca España (Colombia), Bishan Public Library (Singapore), Black Diamond Library (Denmark), Stuttgart City Library (Germany) and Turku City Library (Finland) as well as Bibliotheca Alexandrina (Egypt). Institutional libraries' access is restricted to users who belong to these particular institutions such as the University of Aberdeen New Library (Scotland), Cayenne University Library (French Guiana), Tama Art University Library (Japan), University of Versailles Science Library (France) and Library of the Vienna University of Economics (Austria) as illustrated in Fig. 5.10.

Box 8: Cities on Seaside or Waterfronts Worldwide

In the developed world, 35 of the 40 largest cities are either coastal or situated along a river bank.[2]

5.4 Climate Change Impact on Seaside Cities

Seaside and waterfront buildings including libraries could face serious impacts from severe events caused by climate change. Severe events such as heatwaves, storms and floods as well as landslides have been manifested in many cities during the past decade, particularly Germany, France, Italy, Egypt, the Philippines, India and South Korea.

According to the UN-Habitat's new State of the World's Cities Report 2008/2009 – Harmonious Cities – it indicates that few coastal cities will be spared by climate change.[13] Seaside cities around the world are and will be exposed to climate change risks.

Cities with low-elevation coastal zones around the world are 3351, of which 2145 cities (64%) are located in developing countries. More than half of these cities are in Asia and classified as vulnerable cities, followed by Latin America and the Caribbean (27%) and Africa (15%). In Europe and North America, cities with low-elevation coastal zones are forming two-third and one-fifth of all these cities, respectively.[13] Sea levels rose by an estimate of 17 cm during the twentieth century. Between 1990 and 2080, the global mean projection for sea level rise is estimated to range from 22 to 34 cm. This threat could be alarming especially for low-elevation coastal zone – the continuous area along coastlines that is less than 10 m above sea level presents 2% of the world's land area.[13]

More than half of the 20 largest European cities are developed along river banks which make cities such as Mumbai, Shanghai, Miami, New York City, Alexandria and New Orleans vulnerable to flooding. The risk is nearly the same in Asia as 18 out of the 20 largest Asian cities are either coastal, on a river bank or in a delta which makes cities such as Dhaka, Kolkata, Rangoon and Hai Phong vulnerable to flooding. In addition, African coastal cities such as Abidjan, Accra, Alexandria, Algiers, Cape Town, Casablanca, Dakar, Dar es Salaam, Djibouti, Durban, Freetown, Lagos, Libreville and Lome are also vulnerable to flooding.[13]

Box 9: Climate Change Adaptation
Infrastructure including buildings must be retrofitted to adapt to climate change risks.

According to the Environmental Protection Agency (EPA), climate change has several impacts on seaside cities which include two main effects:

- *The rise in sea level* which leads to the flooding of coastal wetlands, increasing the salinity of ground water and pushing salt water into fresh water bodies.[14] The IPCC predicted an accelerated rise in sea level of 0.2–0.6 m or more by 2100.[15]

---

[13] State of the World's Cities Report 2008/2009: Harmonious Cities, UN-HABITAT, SOWC/08/PR4, 2008. Available at https://unhabitat.org/books/state-of-the-worlds-cities-20082009-harmonious-cities-2/.-. Accessed on March 4, 2017

[14] Climate Impacts on Coastal Areas. Available at https://www.epa.gov/climate-impacts/climate-impacts-coastal-areas. Accessed on March 10, 2017

[15] IPCC, 2007: *Climate Change 2007: Impacts, Adaptation and Vulnerability. Contribution of Working Group II to the Fourth Assessment Report of the Intergovernmental Panel on Climate*

- *The change in the frequency and intensity of storms and precipitation* which severely affects areas with low-quality infrastructure as it may damage properties, disrupts transportation systems, destroys habitats and creates a threat to human health and safety.[14]

1	The increase in ocean temperatures, which in turn will disrupt the existence of marine ecosystems[14].
2	The increase in the acidity of oceans due to absorption of the atmospheric carbon dioxide (CO_2) which may lead to the loss of marine fauna and coral reefs[14].

In light of the above and to lower the climate change risks, cities worldwide need to be more interactive, sustainable and resilient by adopting new techniques for urban infrastructure and management of cities.[14]

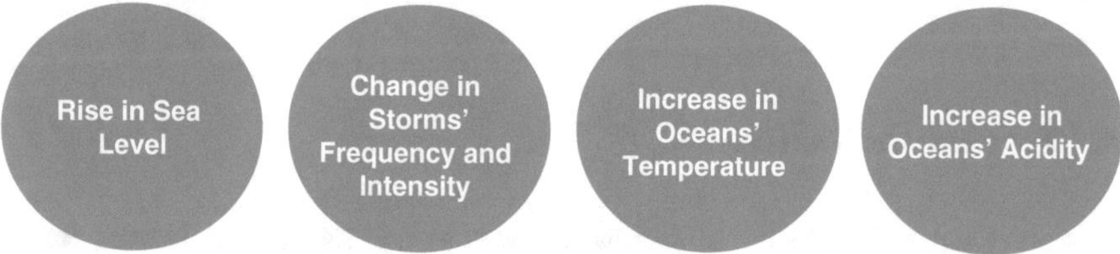

The New Library of Alexandria exhibits many features of sustainability to encounter severe events caused by climate change. (Author)

Image credit: Author

Change, M.L. Parry, O.F. Canziani, J.P. Palutikof, P.J. van der Linden and C.E. Hanson, Eds., Cambridge University Press, Cambridge, UK, 976 pp

5.5 The New Library of Alexandria, Egypt, and Climate Change

The New Library of Alexandria, which is located in the coastal city of Alexandria, is approximately 180 m away from the Mediterranean. Alexandria city is one of the oldest cities on the Mediterranean coast, and it has a 60 Km waterfront stretched from Abu-Qir Bay (east) to Sidi Krier (west).[16] It is the second largest city in Egypt, with a population of about 4 million[17] and a great economic value. The New Library's site is a tourist attraction spot in a unique heritage location. In addition, "the reconstruction of The New Library of Alexandria, attracts one million visitors annually, and is positioning the city as a regional hub of knowledge, science and dialogue among cultures and civilizations".[18] Figure 5.11 shows the external view of the New Library of Alexandria which is classified as a public library. The number of visitors is approximately between 300,000 and 500,000 people annually, for example, it was visited by guided tours amounting to 350,000–400,000 visitors in 2010 (Fig. 5.12). The library' staff is 2500 people.

5.5.1 *Climate Change*

Climate change is already impacting cities that are located on coastal areas. According to the UN-Habitat's report published in 2008/2009, Alexandria city is one of the most vulnerable African cities to rising sea levels.[13] A sea level rise of 0.5 m in Alexandria will lead to the migration of more than 1.5 million people from their homes; loss of 195,000 jobs; monetary loss of about 35 billion US Dollars in industry, tourism and agriculture income; and the irreplaceable loss of famous heritage and cultural sites.[19] Another climate change scenario study stated that eventually Alexandria city would become a peninsula, surrounded by the Mediterranean and is to be accessed only by bridges.[20] As a result of these consequences, the New Library of Alexandria is endangered by rising sea levels.[21]

---

[16] Impact of Climate Change on Egypt. Available at http://www.ess.co.at/GAIA/CASES/EGY/impact.html. Accessed on March 10, 2017

[17] Egypt Population 2017. Available at http://worldpopulationreview.com/countries/egypt-population/. Accessed on March 10, 2017

[18] The Alexandria City Development Strategy, World Bank, 2007. Available at http://siteresources.worldbank.org/INTEGYPT/Resources/Documentation_Alexandria_CDS_June1_2007.pdf. Accessed on March 10, 2017

[19] El-Raey, M., Dewidar, Kh. and El Hattab, M. 1999. Adaptations to the impacts of sea level rise in Egypt. *Climate Research.* 27 August 1999, pp. 117–128

[20] Turner R.K., Kelly P.M. and Kay R.: Cities at Risk, BNA International, London, 1990, as quoted in Hardoy, J. E., Mitlin, D., & Satterthwaite, D. (2013). *Environmental problems in an urbanizing world: finding solutions in cities in Africa, Asia and Latin America*. Routledge

[21] Elsharkawy H., Rashed H., & Rached I., The impacts of Sea Level Rise on Egypt, 45th ISOCARP Congress, 2009

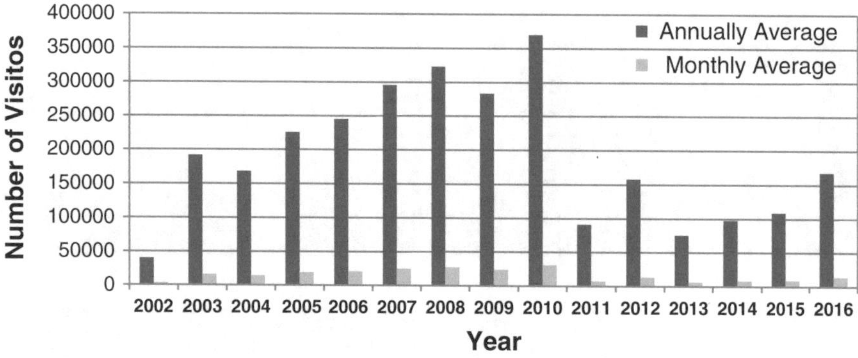

Fig. 5.11 Visitors entered the New Library of Alexandria between 2002 and 2016

Fig. 5.12 External views of the Library of Alexandria, Egypt. (**a**) Library overviewing Mediterranean. (**b**) The main library building. (**c**) South view of the main library hall (Images source: Library of Alexandria (www.bibalex.org))

Fig. 5.13 Severe events manifested by climate change in the city of Alexandria (2015, 2016), Egypt. (**a**) Severe floods in January 2016. (**b**) Traffic is immobilised as a result of severe floods. (**c**) Floods covering the promenade, October 24, 2015. (**d**) Vehicles stuck in the floods in the main street (Images source: www.google.com)

Between 2014 and 2016, the city of Alexandria, Egypt, experienced many severe events that have been manifested by unprecedented torrential rain, which led to dangerous floods. These floods submerged the main thoroughfare in front of the Library of Alexandria giving serious alarms about climate change risks on the city's infrastructure (Fig. 5.13).

5.6 The Case Study: *The New Library of Alexandria (Bibliotheca Alexandrina)*

5.6.1 Introduction and Historical Background

The city of Alexandria has few Greek, Roman, Mamluk and Christian remains. It is now more of a mixture of modern and art deco styles. The ancient Library of Alexandria was the largest library in the ancient world as it provided reference works for great philosophers and writers of the era, including Homer, Plato, Socrates and many more, and a centre for culture and heritage. It is believed that the ancient library was destroyed in a huge fire around 2000 years ago, which resulted in a tragic loss of knowledge and literature.[22]

Around 330 BC, Alexandria city was founded by Alexander the Great. After his death in 323 BC, Alexandria city was transformed from a small fishing village on the Nile delta to becoming a great cultural centre through the Ptolemaic rule of Egypt. The ancient Library of Alexandria was founded around 295 BC.[22]

The main interpretation for the destruction of the ancient Library of Alexandria is that it occurred in 48 BC when Julius Caesar found himself locked in the Royal Palace, surrounded by the Egyptian fleet in the harbour. In order to rescue himself, he set fire to the Egyptian ships, where the fire went out of control and spread to the parts of the city located near the shore, leading to the loss of the library.[22]

The construction of the New Library of Alexandria is reviving the world famous ancient Library of Alexandria. Currently the library is not only considered a heritage revival complex but also a centre for learning in the current digital era. "It acts as a window for Egypt to the outside world". The current library also marks a significant use of sustainability and smart measures which merit our attention.

5.6.2 Public Libraries Design Framework

In developing public libraries, it is vital to understand five main issues to guarantee effective and efficient design and ensure sustainability in the context of sustainable development (SD). Figure 5.14 presents the main sustainability issues and their framework:

A. Context
B. Design features
C. Structure and technology
D. Climate change mitigation techniques
E. Sustainability features

In line with the framework, the following section will describe each element in details (Fig. 5.14).

[22] Haughton, B. (2011, February 01). What happened to the Great Library at Alexandria? Available at http://www.ancient.eu/article/207/. Accessed on March 16, 2017

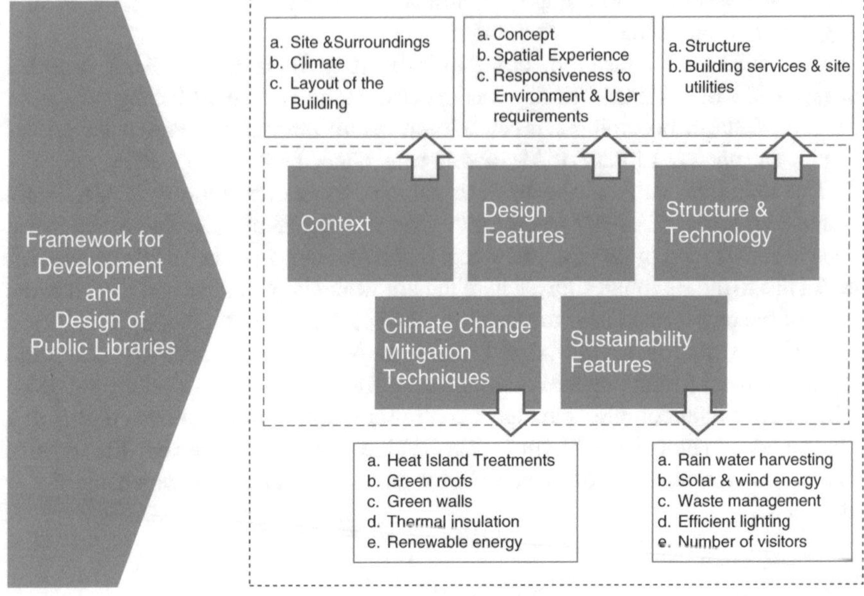

Fig. 5.14 Framework for public library design (Image credit: Author)

5.6.2.1 Context

Site and Surroundings

Climate

In developing public libraries, it is important to understand the climatic characteristics of the city where the library is located. The climate in Alexandria can be classified into two main seasons: mild winter and summer. The mild winter extends from November to April with a minimum temperature 7 °C and averages between 13 and 18 °C. The summer season extends from May to October with temperatures ranging from 21° to a maximum of 38 °C in July and August, with a maximum relative humidity of 95%. The prevailing wind directions are the north and north-west with sand-bearing winds coming from the west in spring.[23] It is clear that the design of the library buildings and the external cladding was developed to reflect this climatic condition and adapt to it.

[23] Bibliotheca Alexandrina. Alamuddin, H., The Aga Khan Award for Architecture 2004. Available at http://www.akdn.org/sites/akdn/files/media/documents/AKAA%20press%20kits/2004%20AKAA/Bibliotheca%20Alexandrina%20-%20Egypt.pdf. Accessed on March 16, 2017

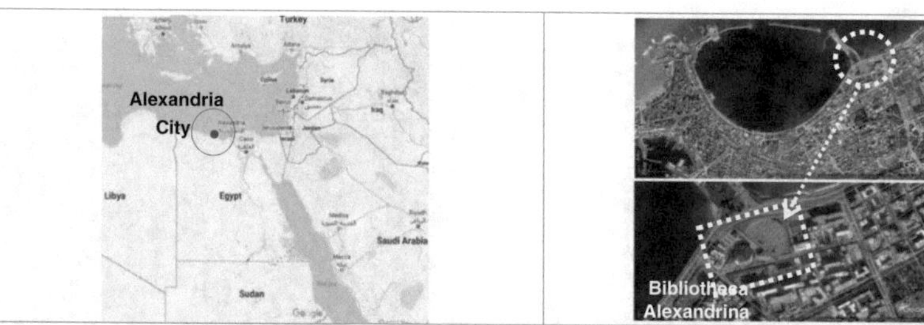

The Bibliotheca Alexandrina, located on a remarkable site on the Eastern Harbor, facing the sea on the North and the Alexandria University. From the south, it is located on Port Said Street[23]. The library is considered a monumental building with great symbols and meanings[24]. According to the 1993 archaeological survey, the library is placed close to the location of the old library in the Brucheion (the Ancient Royal Quarter) [23].

Fig. 5.15 Location of the New Library of Alexandria, City of Alexandria, Egypt (Source: Google maps)

Layout of the Building

Figure 5.15 illustrates the location of the new library in Alexandria. The building's architecture is a result of the collaboration between Snohetta A.S. (Norway) and Hamza Associates (Egypt).[24] Figure 5.16 illustrates the layout (45,000 m^2) which consists of three main buildings: the pre-existing conference centre, the planetarium and the new building. These buildings are connected through an underground level below the plaza into one large functional complex. An external plaza (8500 m^2) is inviting and provides openness to the surroundings.[24] The layout configuration is very suitable, as the New Library circular building is in connection with the smaller sphere of the planetarium (Fig. 5.16). The existing conference centre acts as a counterpoint in the overall massing of the forms.[24] The whole building is surrounded by a water pool which has three roles: (1) reflecting water of the sea on the building façade from various angles, (2) separating the building to protect from threats and (3) giving a feeling of "floatation" to the building, distancing it out of the reach of the surroundings contributing to its uniqueness.[24]

Box 10: The New Library Site Components
The project site has the following spaces:

1. Planetarium (99 seats)
2. Conference hall, 3020 square metres
3. Outdoor plaza, 8500 square metres
4. Reflection water pool, 4600 square metres

[24] A landmark building: reflections on the architecture of the Bibliotheca Alexandrina. Ismail Serageldin – Alexandria: Bibliotheca Alexandrina, 2007. Available at https://www.bibalex.org/Attachments/Publications/Files/LandMark_English.pdf. Accessed on March 16, 2017

a.

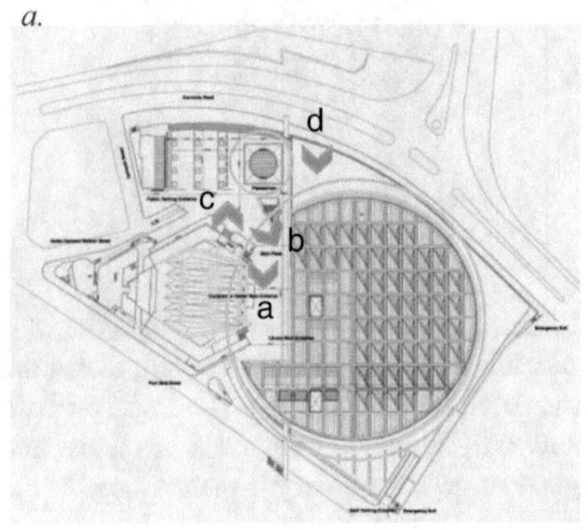

Image source: Library of Alexandria (www.bibalex.org)

b.

c.

d.

e.

Fig. 5.16 The New Library Complex. (**a**) Layout of the New Library of Alexandria (Image source: Library of Alexandria (www.bibalex.org)). (**b**) The pre-existing conference centre, *right.* (**c**) Exterior view of the bridge linking the main reading hall to the outside. (**d**) Exterior view of the planetarium. (**e**) The roof of the reading hall (Images credit: Author)

Box 11: The New Library and Surrounding Environment

The project's design respects the surrounding environment and its location causing no visual barrier for the seaside location.

a.

b.

Fig. 5.17 Exterior of the New Library of Alexandria. (**a**) Library's inclined roof facing north and the Mediterranean. (**b**) A close image of inclined roof of the reading hall (Images credit: Author)

5.6.2.2 Design Features

Concept

The fundamental idea of the building massing is in the form of a disc ascending from the water, which represents the past, tilting towards the future (Fig. 5.17b). This inclined form is facing the prevailing wind and inclined to provide daylighting throughout the day. A granite wall that is carved with letters from alphabets around the world clads the building's facade (Fig. 5.17b). The complex includes the main library, a planetarium and a pedestrian bridge that cuts across the public plaza, which is shared with the existing building of the conference centre as shown in Fig. 5.17a. The image of the library's exterior is created by the elegant slanted roof oriented towards the sea (Fig. 5.17b). It symbolizes the rising of the sun which has influenced ancient Egyptian civilization. The building respects its function through its aesthetically appealing form, spaces created and materials and colours used.[24] In the design of the library, the human scale was taken into consideration, which is presented in two main ways: (1) reducing the building volume by the inclined roof and embedding four floors below ground and (2) splitting the reading areas into seven levels.[24]

Spatial Experience

The New Library of Alexandria consists of 11 floors: 4 below the ground and 7 above. The building is 33 m high. The ground floor is 23,900 square metres and the total floor area is 85,405 square metres. It comprises two main parts: the library and the planetarium which are linked at basement level.[23] Nonetheless, the library's spaces are listed as follows:

- Main reading hall of 13,625 square metres (Fig. 5.18)

Fig. 5.18 The interior of the main reading hall (Images credit: Author)

- Six specialist libraries: the children's library, young people's library, multimedia library, Taha Hussein library for the visually impaired, microfilm and special collections and rare books and manuscripts (Fig. 5.19)
- Four museums, including antiquities, manuscripts and the history of science
- An Internet archive
- Seven research centres
- Fifteen permanent galleries and the Nobel Peace Hall
- Exploratorium
- Supercomputer
- Offices, services and facilities supporting the Library

Daylighting is one of the prime design features in libraries: The New Library of Alexandria is an excellent, exemplary building of daylight provisioning and ensuring environmental sustainability. (Author)

a.

b.

c.

Fig. 5.19 Special libraries. (**a**) Taha Hussein Library for the visually impaired. (**b**) Microfilm and special collections. (**c**) Rare books and manuscripts (Images credit: Author)

Image credit: BA

Responsiveness to Environment and User Requirements

The main reading hall is situated at the eastern sector of the disc. The entrance to the Library, its administrative area, specialized libraries, museums and the other facilities and services are organized on the western sector. The glazing of the circle (the roof of the main reading hall) allows for natural lighting for the administration area, along with the presence of two light wells.[23] Figure 5.20 shows such ceiling's openings.

5.6.2.3 Structure and Technology

Structure

In the design of the library, the main challenge facing the structure was the site's location, as it is close to the sea, especially with having underground levels. The circular shape (the roof of the main reading hall) was used to limit water penetration instead of expansion joints in the diaphragm wall. The existing conference hall, along with the varying temperature conditions and the diaphragm wall's length, has put a major constraint on the design of such wall. It was analysed and studied using computer modelling and programmes. Computer technology was also used in the design of the form of the building as the shape was not a cylinder but a "section form called a torus".[23]

The main structural system of the Library comprises a circular wall with a concrete column grid of 7.2 × 9.6 m for the lower floors and 14.4 × 9.6 m for the floors

a. *b.*

c.

Fig. 5.20 The main reading hall's ceiling. (**a**) The ceiling of the main reading hall. (**b**) The ceiling with openings. (**c**) A view of a set of openings (Images credit: Author)

above ground level. The whole is carried on a raft foundation with a pile for each column to reduce the construction time and cost.[23]

The planetarium has a steel structure that is suspended from a steel bridge which spans the diaphragm wall that in turn is supported on a raft foundation. The pedestrian bridge has an independent steel structure.[23]

Building Services and Site Utilities

The building is connected to the city's main services of water, power and sewerage. It has two electricity feeders as well as three generators and nine UPS (uninterrupted power systems) batteries for emergency use.[23]

5.6.2.4 Climate Change Mitigation Techniques

Climate change mitigation is a necessary action, not only on the scale of cities' resilience but also on the scale of buildings. For seaside libraries, there are many challenges for the designer to overcome the climate change risks such as floods' impact and the increase of the water salinity. Figure 5.21 lists the techniques used

Sandstorm, dust, wind	Flood	High Temperatures	Saline Water	Earthquakes
• Building Form • Facade Material (Granite) • Water Pool	• Special Foundation	• Heat Capacity Material • Water Pool	• Facade Material (Granite)	• Walls covered with 50 CEN of high quality concrete

Fig. 5.21 Climate change mitigation techniques in the New Library of Alexandria (Chart credit: Author)

for climate change mitigation including renewable energy which is extensively presented in Sect. 5.6.1.5: Solar Energy Use, pages 35–36.

When designing libraries in seaside cities, it is important to consider the impact of climate change. In this context, heat island impact has to be properly and effectively treated. In order to offset the heat island effect (HIE), it is vital to treat buildings' roofs, select proper materials and increase green areas and water ponds.

Heat Island Treatments

- Select bright colour materials for pavements such as streets' asphalt and parking areas (i.e. with White Portland Cement – WPC – with high solar reflective index, SRI = 86%).
- Treat the building's roof to incorporate green roof and, if possible, green walls. In the New Library of Alexandria, green walls and green roofs were not incorporated.
- Increase vegetation areas and make water ponds available on the site.

Water Pond Availability

The New Library of Alexandria has exhibited features and treatments of offsetting heat island effect. In this context, water ponds were made as part of the design of the building site to lower the air temperature due to evaporative cooling effect. Figure 5.22 below presents images of the water pond near the planetarium and the main reading hall.

Thermal Insulation Materials

In order to stabilize temperature, exposed concrete (using thermal capacity) was used in the structure of libraries, which provided a balance between the envelop of the building, lighting levels and associated environmental systems.[25] Also, the external walls and cladding of the New Library of Alexandria show that such walls have excellent thermal insulation, with U-value 0.029 W/m^2 K° as shown in Fig. 5.23.

[25] Library Trends, Vol. 60, No. 1, 2011 ("Library Design: From Past to Present," edited by Alistair Black and Nan Dahlkild), pp. 190–214. 2011 The Board of Trustees, University of Illinois

a.

b.

c.

Fig. 5.22 Offsetting heat island effect by using water ponds and bright colour material. (**a**) Water pond near the main reading hall. (**b**) A close view of the water pond. (**c**) Bright colour pavements (Images credit: (**a**, **b**) BA and (**c**) Author)

The wall insulation, made of five layers with air gaps (Fig. 5.23c), showed the effect of the insulation with two air gaps of 20 and 60 cm, on delaying the heat transfer into the main reading hall, hence reducing the cooling loads.

The external walls of the library as shown in Fig. 5.23 have many layers to control the heat flow from outside to the inside of the building. The description of such wall (Fig. 5.23b, c) is as follows:

- The exterior granite panels are 20 cm thick; this is a curving cliff face of roughly cut grey Schulman granite quarried in Egypt – South of Aswan.
- Air gap average 20 cm.
- The reinforced concrete internal wall is a partially sunken cylinder that creates a circular plan shape and supports etched granite cladding panels by thickness 60 cm.
- Gap between the wall and the panels is approximately 60 cm.
- The precast panels are all concrete with sound attenuation holes cast in regular intervals, to an individual radius to fit the perimeter wall of the building about 20 cm[26] (Table 5.2).

[26] Library of Alexandria

a.

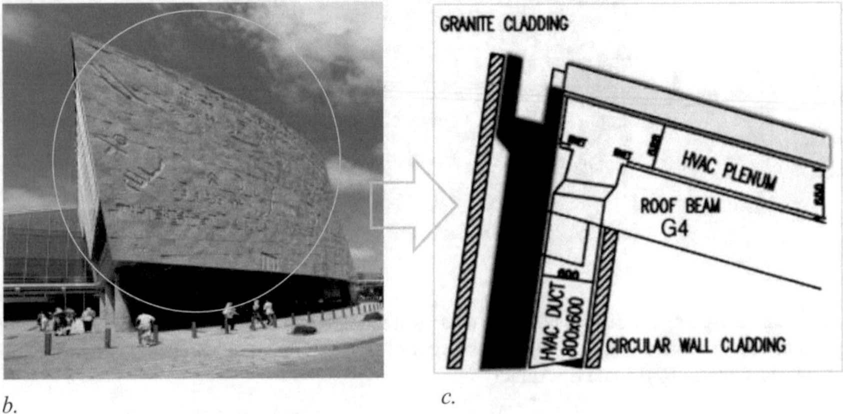

b.

c.

Fig. 5.23 The thermal insulation of external wall – main reading hall. (**a**) Granite cladding of external walls. (**b**) External granite cladding. (**c**) Details of the thermal insulation wall layers (Images credit: (**a**) Author and (**b, c**) BA)

The overall U-value of the external wall is calculated based on the thermal resistance of the layers (seven layers including the outside and inside air).

The following is showing the value of the total wall resistance (R-value) and the (U-value).

Total Wall R-value =	0.04 + 0.067 + 8.33 + 0.57 + 25 + 0.2 + 0.13 = 34.337 **m² K/ W**
Total Wall U-value =	$\frac{1}{34.337}$ = 0.029 W/**m² K**

Table 5.2 Wall layer description

Material	Thickness (m)	Thermal conductivity (W/mK)	R-value (m^2K/W)
Air outside	0.02		0.04
Granite panels	0.2	3	0.067
Air gap$_a$	0.2	0.024	8.33
Reinforced concrete	0.6	1.04	0.57
Air gap$_a$	0.6	0.024	25
Concrete precast panels	0.2	1	0.2
Air inside	0.02		0.13

5.6.2.5 Sustainability Features

The New Library of Alexandria provides a great example of achieving sustainability and a high-quality interior environment in library buildings without high rate of energy demands. This was accomplished by giving attention to major issues such daylighting, lighting, natural ventilation and thermal capacity.[15]

- A summary of the sustainability measures and energy management techniques used in the process is shown in Fig. 5.24.

The provision of daylighting inside the main reading halls in libraries – without allowing glare, to achieve the optimum daylight factor (DF) – is one of the crucial elements of efficient design. (Author)

Image credit: Author

Daylighting Enhancement

Light wells introduce proper natural lighting except for the underground floors which have no natural lighting. Natural lighting is well used in the main reading hall. There was accurate computer-based design for the natural light impinging through the roof panels. Glazing surfaces were designed in a way that does not cause glare. Figure 5.25 shows the special features and design details to reduce glare. Daylighting was allowed in some sections of the Library's main building, where lighting enters the special collection hall through the courtyard using a special type of glass (Fig. 5.26).

Indoor Sustainability Measures	Energy Management Techniques
• **Daylighting enhancement** • **Noise reduction using insulating Indoor acoustic materials** • **Rain water harvesting, reuse water for cleaning and irrigation** • **Waste Management (Paper recycling)** • **Building Management System (BMS) coordination:** • indoor temperature, • fire system, • security system and • air conditioning • **Efficient Glass** • **Durable material requiring little maintenance**	• **Natural lighting** • **Using LED lighting and reducing conventional lighting units by 20 %.** • **Using lighting dimmers in all halls.** • **Air Conditioning Management via BMS** • **Using solar energy (PV panels)**

Fig. 5.24 Sustainability features in the New Library of Alexandria (Chart credit: Author)

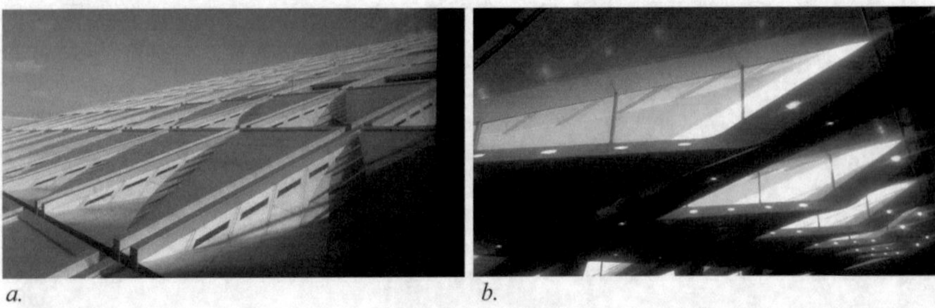

a. *b.*

Fig. 5.25 Natural daylighting in the New Library of Alexandria. (**a**) Special openings, main reading hall (outside). (**b**) Opening detail for daylighting (inside) (Images credit: Author)

a. b.

Fig. 5.26 Lighting entering the library's main special collection section. (**a**) Daylight provision via the courtyard. (**b**) Special glass to allow for daylight without glare (Images credit: BA)

Noise Reduction Using Indoor Acoustic Materials

The library building has excellent acoustics and noise insulation, especially in the main reading hall as shown in Fig. 5.27 below. The concrete panels have many holes which act as a sound absorbent (pigeon hole). These air pockets were invented during the Ancient Egyptian era to store the papyrus papers. The acoustic wall covers the whole internal walls from level 7 to level 0 (Fig. 5.27a, b). In the circulation area, where large numbers of users are moving up and down from the staircases and the elevators, copper panels clad the vertical walls to offset noise by absorbing the sound waves and preventing resonance of sound. In addition, concrete, aluminium and timber panels, which have acoustic slots, are used for acoustic insulation.

Rainwater Harvesting

The New Library of Alexandria also incorporates a system of rainwater harvesting, which annually saves about 3.4% of the water consumption (annual water consumption in the main building is 76,000 m^3 while the annual rainwater usage is 2,550 m^3).

Fig. 5.27 Noise reduction using indoor acoustic materials. (a) Internal acoustic wall for sound absorption. (b) Part of the acoustic wall at the maps reading area. (c) Copper cladded panels, circulation lobby (Images credit: Author)

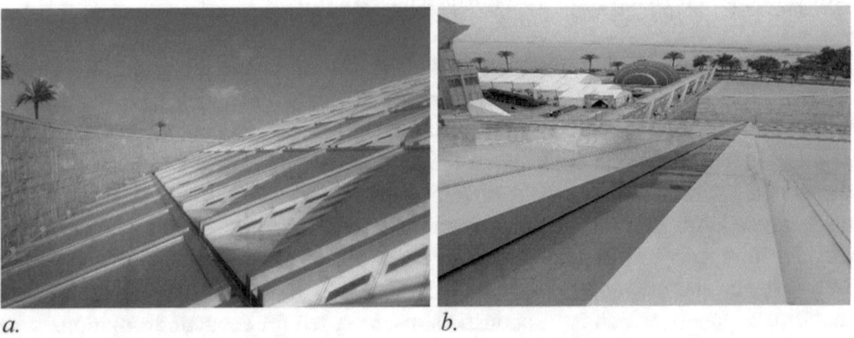

Fig. 5.28 Rainwater harvesting and treatment. (a) The roof of the library main reading hall. (b) Rain water rolling – roof facing the planetarium (Images credit: Author)

Rainwater is collected from the roof and filtered to be used again for cleaning purposes. This is one of the sustainability measures that was manifested in the library operation. Figure 5.28 shows the roof's rainwater harvesting system on the roof of the main reading hall.

Waste Management

Waste recycling is used in the daily operation of the New Library of Alexandria to exhibit one of the important sustainability measures. The Library of Alexandria developed a waste management strategy of paper recycling that makes use of large amounts of waste paper and properly disposing them to be recycled. Since July 2014, an amount of 24,990 Kg of paper and cardboard were collected for recycling. The BA waste management is a project managed by the Centre for Environmental Studies (CES). Its task force comprises of volunteers from university students that is part of the youth capacity building programmes. The paper is collected from offices in the BA building, conference hall building and rented apartments that are belonging to the BA. This process is called "source sorting". Figure 5.29 below illustrates the process of waste management.[27]

Building Management System (BMS) Coordination for Indoor Temperature, Fire System, Security System and Air-Conditioning

Building management system (BMS) is a computer-based system that controls and monitors the building's systems including access and security, lighting, HVAC (heating, ventilating and air-conditioning) and firefighting to ensure the building operates at maximum levels of efficiency. In the Library of Alexandria, indoor temperature levels are kept in comfortable levels by managing the building's air-conditioning through BMS. BMS also controls fire and security systems in case of any emergency situation. The building has three generators and nine UPS which are used for the emergency lighting and security access systems in case of power cuts. In case of an emergency, the monitors and security cameras installed in the building provide the users with means of egress out of the building. The main reading hall can be evacuated in 15 min. The lighting and air-conditioning systems installed in the ceiling are shown in Fig. 5.30.

Efficient Glass

Installing high-performance glass on building facades is a necessity. Efficient glass can save 90 million tons of CO_2 every year in Europe, which is equivalent to the emissions of 9.8 million Europeans over 1 year. The following characteristics are

Fig. 5.29 Rainwater harvesting and treatment (Image credit: Author)

[27] Meeting at the Library of Alexandria dated on March 13, 2017

a. b.

Fig. 5.30 Lighting and air-conditioning systems installed in the ceiling. (**a**) Conventional artificial lighting units. (**b**) HVAC duct in ceiling (Image credit: Author)

the determinants of the glass efficiency,[28] and Table 5.3 presents the type of glass used in the New Library of Alexandria.

- Light transmission (LT)
- Shading coefficient (SC)
- Solar reflection (SR)
- Total solar radiant heat transmission factor (SHGF)
- UV reduction factor and U-value

Durable Material Requiring Little Maintenance

The materials used in the New Library of Alexandria have been selected mainly from natural resources to ensure sustainability and durability. This has been integrated in many areas of the main reading hall and the external walls and pavements such as granite, marble, wood, glass, concrete and others. Materials chosen for the project are durable and require little maintenance. For example, the granite cladding of the external wall and the aluminium cladding of the roof of the main reading hall are great examples to ensure sustainability (Fig. 5.31).

Using LED Lighting

Lighting is forming 30% of the total electrical energy use. To reduce energy consumption, the library management replaced 20% of the conventional lighting units with LED in the main building (6000 LED) and about 90% of the conference centre

[28] High performance glazing for a sustainable habitat, Saint-Gobain glass. Available at http://nordic.saint-gobain-glass.com/environment/pdf/high-performance-glazing-for-sustainable-habitat.pdf. Accessed on March 26, 2017

Table 5.3 Efficient solar glass properties

Glass type A	Criteria	Required	Achieved	Building and location
GT.3 3A/1 Double glazed Pane	Light transmission (LT)	Min. 50%	62–33%	Inclined walls to the perimeter of the planetarium 
	Shading coefficient (SC)	< 45%	44–30%	
	Solar reflection (SR)	Max. 10%	9–27%	
	Total solar radiant heat transmission factor (SHGF)	Max. 40%	38–26%	
	UV reduction factor	NA	99.5%	
	U-value	1.35 W/m² K°	1.45 W/m² K°	

Glass type B	Criteria	Required	Achieved	Building and location
GT.1 Double glazed Pane	Light transmission (LT)	Min. 50%	62%	Vertical glass roof provides light to the diagonal truss, roof panel type A 
	Shading coefficient (SC)	< 45%	44%	
	Solar reflection (SR)	Max. 10%	9%	
	Total solar radiant heat transmission factor (SHGF)	Max. 40%	38%	
	UV reduction factor	Min 95%	99.5%	
	U-value	1.35 W/m² K°	1.35 W/m² K°	

Glass type C	Criteria	Required	Achieved	Building and location
GT.3D/1 Double glazed Pane	Light transmission (LT)	Min. 50%	62%	Inclined and vertical glass walls provide light to courtyards No.: 1, 3 Vertical glass walls provide light to courtyards No.: 5, 6 West entrance wall East basement wall inclined north wall in front of the café 
	Shading coefficient (SC)	< 45%	43%	
	Solar reflection (SR)	Max. 10%	9%	
	Total solar radiant heat transmission factor (SHGF)	Max. 40%	38%	
	UV reduction factor	NA	–	
	U-value	1.35 W/m² K°	1.4 W/m² K°	

(continued)

Table 5.3 (continued)

Glass type D	Criteria	Required	Achieved	Building and location
GT.2A Double glazed Pane	Light transmission (LT)	Min. 50%	33%	Sloping glass roof lights in the plan of the roof, roof panel type D and F
	Shading coefficient (SC)	< 45%	33%	
	Solar reflection (SR)	Max. 10%	27%	
	Total solar radiant heat transmission factor (SHGF)	Max. 40%	26%	
	UV reduction factor	Min 95%	99.5%	
	U-value	1.35 W/m^2 K°	1.3 W/m^2 K°	

Glass type E	Criteria	Required	Achieved	Building and location
GT.3E Double glazed Pane	Light transmission (LT)	Min. 50%	64%	Sliding doors for balconics 1–15 and adjacent fix glasses
	Shading coefficient (SC)	< 45%	44%	
	Solar reflection (SR)	Max. 10%	9%	
	Total solar radiant heat transmission factor (SHGF)	Max. 40%	39%	
	UV reduction factor	NA	–	
	U-value	1.35 Wlm^2 K°	1.4 W/m^2 K°	

Glass type F	Criteria	Required	Achieved	Building and location
GT.6A/2–50 Single toughened Glass	Light transmission (LT)	–	60%	Glass shade of the café terrace
	Shading coefficient (SC)	–	67%	
	Solar reflection (SR)	–	26%	
	Total solar radiant heat transmission factor (SHGF)	–	59%	
	UV reduction factor	–	–	
	U-value	–	5.8 W/m^2 K°	

(continued)

Table 5.3 (continued)

Glass type G	Criteria	Required	Achieved	Building and location
GT.3C Double glazed Pane	Light transmission (LT)	Min. 50%	65%	Curved glass walls provide light to circular court No. 4
	Shading coefficient (SC)	< 45%	66%	
	Solar reflection (SR)	Max. 10%	18%	
	Total solar radiant heat transmission factor (SHGF)	Max. 40%	57%	
	UV reduction factor	NA	–	
	U-value	1.35 W/m² K°	1.6 W/m² K°	

Source: BA

a. Gray Schulman Granite quarried in South Aswan, Egypt

b. Bright coloured materials

c. Granite and marble with bright colour –Local Materials

Fig. 5.31 Natural materials applied throughout the building to achieve sustainability. (**a**) Grey Schulman granite quarried in South Aswan, Egypt. (**b**) Bright coloured materials. (**c**) Granite and marble with bright colour – local materials (Images credit: Author)

Fig. 5.32 Efficient lighting – LED lamps in the main building. (a) Conference centre with LED lamps. (b) LED lamps in the main entrance. (c) An office, in the main building with replaced LED (Images credit: BA)

in order to save energy. Figure 5.32 shows the spaces that have LED lamps substituting the inefficient lamps including the conference centre, storages, the main entrance (Fig. 5.32b), and some offices (Fig. 5.32c). The total saving from efficient lighting is amounting to 2221207.6 kWh and the percentage of saving is 77%. This led to CO_2 emission reduction of 1199 tCO_2 of the total emissions.

Solar Energy Use

The New Library of Alexandria is a pioneer organization when it comes to renewable energy. It tries to be updated with the latest sustainability trends; thus it develops ways to move forward in becoming a greener institution. Although the initial design of the library didn't incorporate the use of renewable energy, a system of photovoltaic (PV) panels was installed in the library to generate electrical energy as shown in Fig. 5.33 below.

The installation of the solar PV panels is connected to the grid with 10 kW power maximum. The total number of the PV panels is 46; each panel capacity is 250 watt. Due to the accumulated dust in summer, starting from March, the power output of the 46 panels is reduced to 5.3 kW power on March 13, 2017. The power production of 48 kW (till that moment) is enough for the great hall.

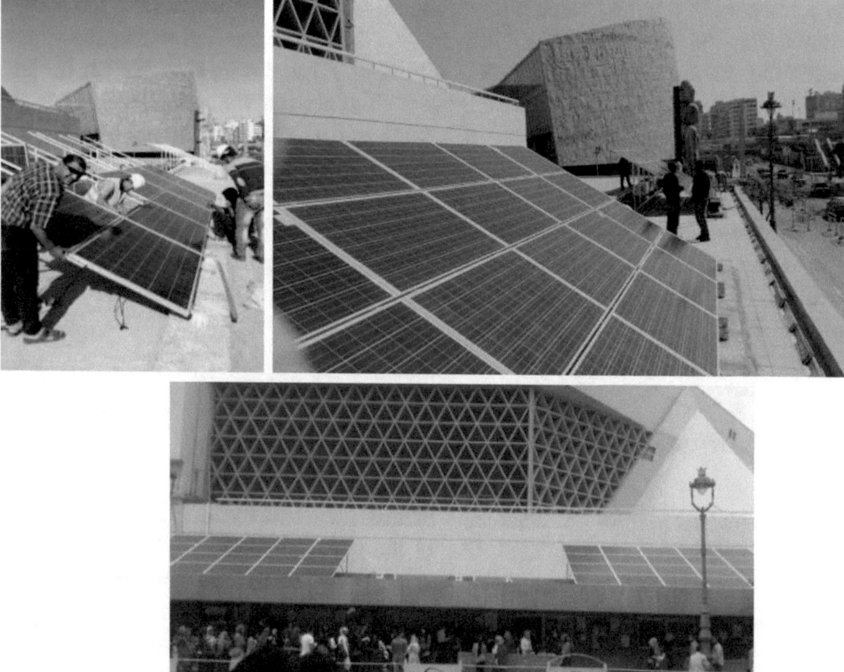

Fig. 5.33 Solar photovoltaic panels on the roof of the Library of Alexandria (Image source: BA)

5.7 Building's Structure Assessment

The library building has been designed to withstand food impact and earthquakes. A thick concrete wall has been erected underneath the foundation of the library to prevent any underground water or flood impact to penetrate to the foundation of the whole building. This has been done in the early construction of the library. This technology of underground water prevention is seen as one of the sustainability measures to reduce the cost of maintenance. The structure assessment has been outlined in two major points:

- The technology, with the exception of the foundations and structure, is straightforward for any large building.
- The effect of earthquakes was checked using computer-generated models.

It has been noticed that readers' access for staff is connected to fan coil and is turned off when they leave the space. Also, there is no solar water heating to heat the water in the library, but all water heaters are electric. In addition, the BMS system is connected to the fan coil as a time programme, e.g. once it is 19:00, the light is turned off. The BMS, which is in the process of being upgraded to Siemens, measures the comfort level, relative humidity (RH %) and temperature. For the air-conditioning (AC) in the main building, six chillers (four water cool and two using air cooling) are installed. In terms of operation, four chillers (two are on duty and two on standby) and two chillers (one in duty and a second is on standby) are installed. In this way, energy is saved. Moreover, two variable speed drives (VSD) were installed in 2009 to make the AC system more efficient. Only one chiller is active after 19:00 to maintain relative humidity and temperature inside the library maintained at certain level and keep essential systems operating such as server and main frame.

5.8 Conclusions

In this chapter, the enormous challenges that face seaside cities and waterfront buildings are introduced and discussed. One of the major challenges is how these cities adapt to severe events caused by climate change impacts such as heat waves, storms and floods, which have been manifested in cities during the past decade. Also, the chapter examined seaside public libraries worldwide by analysing three libraries located in different climatic zones: the John F. Kennedy Presidential Library in Boston, Massachusetts, USA; the Royal Danish Library in Copenhagen, Denmark, Europe; and the Library at the Dock in Docklands, Melbourne, Australia. The library in Boston shows no sign of sustainability per say, but the library in Melbourne, Australia (which is constructed with cost-effective environmentally friendly materials), includes many sustainability features. It is considered economically sustainable, socially sustainable and environmentally friendly, and it acts as a social hub. Nevertheless, the library in Copenhagen, Denmark has fewer sustainability features when compared with that of Melbourne which addressed sustainability in terms of provision of natural daylighting and is socially sustainable. This is basically due to the fact that the library in Boston was built in 1979, i.e. before the sustainability era.

The types of public buildings were also classified with reference to library types. In order to develop and design a public library, a five-step framework was developed and adopted to guarantee effective and efficient design and ensure sustainability. This framework includes the context, design features, structure and technology, climate change mitigation techniques and sustainability features.

In order to gain further knowledge about the design principles of seaside buildings, emphasis on the New Library of Alexandria in Egypt could be considered as an excellent case study from a seaside public library viewpoint, to inform and assist in new libraries development. This Library was analysed and assessed in terms of sustainability measures including: daylighting enhancement, noise reduction using insulating indoor, acoustic materials, rainwater harvesting, reuse grey water for cleaning and irrigation, as well as waste management (paper recycling), building management system (BMS) coordination (indoor temperature, fire system, security system and air conditioning), efficient glass, and durable material requiring little maintenance.

Based on the observation, site visit and assessment, the New Library of Alexandria has achieved many sustainability principles and a high-quality interior environment, but more action is needed. Nonetheless, the sustainability features and elements in the library are identified and analysed. This highlights that the designer and the library management faced many challenges to bring together the different sustainability requirements in an innovative way.

As far as the lighting is concerned, it forms 30% of the total electrical energy use in the New Library's buildings. To save energy, conventional lighting units were replaced by LED lamps in the main building and about 90% of the conference centre, storages, main entrance and some offices in the main building. As a result, the total saving from efficient lighting is amounting to 2,221,208 kWh and the percentage of saving is 77%. This led to CO_2 emission reduction of 1199 tCO_2 of the total CO_2 emissions.

Also, the New Library of Alexandria is considered a smart organization when it comes to sustainable energy. It integrates the latest sustainability trends and developed innovative ways to become a greener institution. Although the initial design of the library didn't incorporate the use of renewable energy, but many sustainability principles were incorporated, notably an on-gird photovoltaic (PV) system of 46 panels providing a maximum of 10 kWp to generate electrical energy, but due to the dust in summer months, the power output is reduced to 5.3 kWp according to March 13, 2017, recoded data.

Finally, it has been noticed that the New Library of Alexandria demonstrated many features of sustainability and presents an exemplary model of public libraries in a seaside city – Alexandria, Egypt – to counterbalance climate change impact and adapt to its risks.

Acknowledgement The author would like to utter his sincere thanks to *HE Ambassador Dr. Mostafa M. El Feki*, Director General of The New Library of Alexandria, *Dr. Ismail Serageldin*, Immediate past Director General of The New Library of Alexandria (Bibliotheca Alexandrina), and *Eng. Hoda Elmikaty*, Deputy Director of Bibliotheca Alexandrina, for their kind approval to consider the New Library of Alexandria as the case study representing one of the seaside libraries. Special thanks to *Dr. Azza ElKholy*, Head of Academic Research Sector, and *Eng. Tarek Yassin*, Head of Engineering Sector, for their assistance and cooperation. The author also expresses his sincere thanks to *Dr. Marwa Elwakil*, Director Centre for Special Studies and Programs (CSSP), and *Ms. Heidi El-Shafei*, Acting Head of the Special Programs Unit, for their support, the facilitation of the site visit to the BA buildings, data gathering and the drawings of the library as well as their hospitality. The author is also grateful to *Eng. Fagr El Guerzawi*, Director of Electro-mechanical Operation and Maintenance Department, Engineering Sector, and *Eng. Tarek Helmy*, Director of Construction Maintenance Department, Engineering Sector, as well as *Eng. Mohamed Khamis*, Head of Electrical Power Section and the Engineering Team, for the demonstration and description of the BA concerning the sustainability measures such as renewable energy and efficient lighting integrated in the library. In addition, the author would like to voice his appreciation to *Eng. Faten Fares*, Acting Head of Architecture Maintenance Unit, for her efforts and providing the technical data and the information regarding the main building and the images of the library as well verifying all the data and information in the book chapter. Moreover, heartfelt thanks go to *Israa Elfayoumi, Yousra Sobeih, Hanan Mounir, Rola Amin* and *Mohamed Mahmoud* – Bibliotheca Alexandrina – for their coordination and efforts.

Moreover, the author expresses his special thanks to my assistant *Arch. Mona Mostafa* and *Arch. Yasmine Mohamed Nasr* for their support, time, great efforts in conducting the research on the literature and also assisting me throughout the preparation of this extensive work and in developing this chapter, notably gathering information. The author would also like to express his heartfelt appreciation to all the teams who participated in developing this work. Last but not least, the author would like to thank *Arch. Mohamed Gad* for his assistance and effort exhibited during the final review and proofing of this book chapter.

References

About the John F. Kennedy Presidential Library. Available at: https://www.jfklibrary.org/About-Us/About-the-JFK-Library.aspx. Accessed 14 Mar 2017.

About the John F. Kennedy Presidential Library, I.M. Pei, Architect. Available at: https://www.jfklibrary.org/About-Us/About-the-JFK-Library/History/IM-Pei--Architect.aspx. Accessed 14 Mar 2017.

Alamuddin, H. *The Aga Khan Award for Architecture 2004*. Bibliotheca Alexandrina. Available at: http://www.akdn.org/sites/akdn/files/media/documents/AKAA%20press%20kits/2004%20AKAA/Bibliotheca%20Alexandrina%20-%20Egypt.pdf. Accessed 16 Mar 2017.

Black, A., & Dahlkild, N. (Eds.). (2011). Library Design: From Past to Present. *Library Trends, 60*(1), 190–214. The Board of Trustees, University of Illinois.

Black Diamond, Royal Danish Library, Librarybuildings.Info. Available at: http://www.library-buildings.info/denmark/black-diamond-royal-danish-library. Accessed 14 Mar 2017.

Cheng, L. (2014, June 5). The Library at the Dock. *ArchitectureAU*. Available at: http://architectureau.com/articles/the-library-at-the-dock/#img=2. Accessed 14 Mar 2017.

Climate Impacts on Coastal Areas. Available at: https://www.epa.gov/climate-impacts/climate-impacts-coastal-areas. Accessed 10 Mar 2017.

Egypt Population 2017. Available at: http://worldpopulationreview.com/countries/egypt-population/. Accessed 10 Mar 2017.

El-Raey, M., Dewidar, K., & El Hattab, M. (1999, August 27). Adaptations to the impacts of sea level rise in Egypt. *Climate Research*, 117–128.

Elsharkawy, H., Rashed, H., & Rached, I.. (2009). The impacts of Sea Level Rise on Egypt, 45th ISOCARP Congress.

Haughton, B. (2011, February 1). *What happened to the Great Library at Alexandria?* Available at: http://www.ancient.eu/article/207/. Accessed 16 Mar 2017.

High performance glazing for a sustainable habitat, Saint-Gobain glass. Available at: http://nordic.saint-gobain-glass.com/environment/pdf/high-performance-glazing-for-sustainable-habitat.pdf. Accessed 26 Mar 2017.

Impact of Climate Change on Egypt. Available at: http://www.ess.co.at/GAIA/CASES/EGY/impact.html. Accessed 10 Mar 2017.

IPCC. (2007). *Climate change 2007: Impacts, adaptation and vulnerability. Contribution of Working Group II to the Fourth Assessment Report of the Intergovernmental Panel on Climate Change* (976 pp), M. L. Parry, O. F. Canziani, J. P. Palutikof, P. J. van der Linden, & C. E. Hanson (Eds.). Cambridge, UK: Cambridge University Press.

LeMaire, G. (2011, August 7). AD classics: JFK Presidential Library/I.M. Pei. *ArchDaily*. Available at: http://www.archdaily.com/153285/ad-classics-jfk-presidential-library-i-m-pei. Accessed 14 Mar 2017.

Library at the Dock, Claredesign. Available at: http://claredesign.com.au/portfolio/public/docklands-library/docklands-library-info/. Accessed 14 Mar 2017.

Library at the Dock, Hayball Projects. Available at: http://www.hayball.com.au/projects/docklands-library/. Accessed 14 Mar 2017.

List of the top 25 coastal cities in the world. Available at: http://www.cntraveler.com/galleries/2016-06-24/the-25-best-coastal-cities-in-the-world. Accessed 4 Mar 2017.

Meeting at the Library of Alexandria dated March 13, 2017.

Monolithic Modern Design-The Royal Library In Denmark By Shl Architects. Homesthetics.net. Available at: http://homesthetics.net/the-royal-library-in-denmark-by-shl-architects/. Accessed 14 Mar 2017.

Serageldin, I. (2002). *Bibliotheca Alexandrina: The rebirth of the Library of Alexandria*. Alexandria: Bibliotheca Alexandrina.

Serageldin, I. (2007). *A landmark building: reflections on the architecture of the Bibliotheca Alexandrina*. Alexandria: Bibliotheca Alexandrina. Available at: https://www.bibalex.org/Attachments/Publications/Files/LandMark_English.pdf. Accessed 16 Mar 2017.

State of the World's Cities Report 2008/2009: Harmonious Cities, UN-HABITAT, SOWC/08/PR4, 2008. Available at: https://unhabitat.org/books/state-of-the-worlds-cities-20082009-harmonious-cities-2/. Accessed 4 Mar 2017.

The Alexandria City Development Strategy, World Bank, 2007. Available at: http://siteresources.worldbank.org/INTEGYPT/Resources/Documentation_Alexandria_CDS_June1_2007.pdf. Accessed 10 Mar 2017.

The Black Diamond, Arcspace. Available at: http://www.arcspace.com/features/schmidt-hammer-lassen-architects/the-black-diamond/. Accessed 14 Mar 2017.

The Library at the Dock, Arcspace – Available at: http://www.arcspace.com/features/clare-design/the-library-at-the-dock/. Accessed 14 Mar 2017.

The Royal Library, Schmidt Hammer Lassen Architects. Available at: http://www.shl.dk/the-royal-library/. Accessed 14 Mar 2017.

Turner, R. K., Kelly, P. M., & Kay, R. (2013). Cities at risk, BNA International, London, 1990, as quoted in Hardoy, J. E., Mitlin, D., & Satterthwaite, D. In *Environmental problems in an urbanizing world: finding solutions in cities in Africa, Asia and Latin America*. New York: Routledge.

Chapter 6
Seaside Buildings in Portugal

Manuel Correia Guedes, Bruno Marques, and Gustavo Cantuária

6.1 Introduction

Portugal mainland has 1230 Km of coast, almost half of its borders. The human pressure upon the seacoast is a growing phenomenon – recent estimates show that more than 80% of the population will live in these coastal areas in the near future. This human pressure upon the coast is a true challenge in terms of sustainable development (cf. Fig. 6.1). It often leads to the degradation of the living space, and valuable natural resources. Also, global warming is a serious threat for seaside areas, with the increase in frequency and intensity of extreme phenomena such as strong winds and hurricanes, floods, and sea level rise. The following chapters will present an overview on the evolution and present situation of seaside building in Portugal, both at urban and building level, showing good and bad examples, and pointing out best practice design strategies.

6.2 Urban Planning

6.2.1 *Brief Overview*

The tendency to occupy coastal areas has always been a constant throughout history. This can be explained by several factors:

- Transport: until the 1850s, the sea was the best mean of transport of people and cargo – fastest, safest, without any obstacles. The largest cities in Portugal – Lisbon and Oporto – are located by the sea. The name of Oporto itself derives from (sea) port.

M.C. Guedes (✉) • B. Marques
Higher Technical Institute (IST), University of Lisbon, Portugal
e-mail: manuel.guedes@tecnico.ulisboa.pt

G. Cantuária
University of Brasília (UNICEUB), Brasil

A. Sayigh (ed.), *Seaside Building Design: Principles and Practice*,
Innovative Renewable Energy, https://doi.org/10.1007/978-3-319-67949-5_6

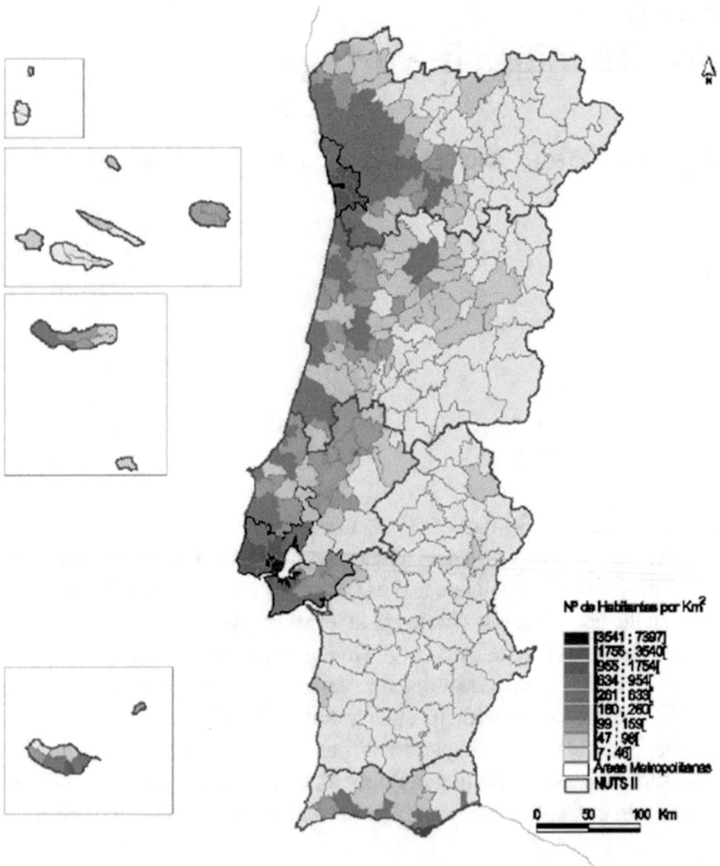

Fig. 6.1 Number of inhabitants per m² in Portugal in 2011 (Source: INE 2014)

- Safety: the interior of the country was more prone to be invaded. There are military fortifications and castles all along the border with Spain.
- Commerce: the main access routes to commerce, first with Flanders ad the Hanseatic League and after the 1500s with Africa, India and Brazil, were through the sea.
- Food: seaside locations allow for fishing – an important source of food. They also offer a more temperate climate, better for agriculture. The interior was never very productive in terms of agriculture.
- Leisure: by the end of the nineteenth century, the urban population began to take an interest on going to the beach. In Lisbon, the first beaches, such as Alcântara, Pedrouços or Algés, were very close to the city centre. Later, with the appearance of the train and trams, new beaches further away from the city became popular, such as Carcavelos, Estoril or Cascais, the same with Oporto, with the beaches of Foz, Granja, Espinho, Figueira da Foz, etc. Today, seaside tourism represents 10% of the country's GDP.

In terms of urban design, coastal cities grew slowly for centuries, until the rural migrations brought about by the industrial revolution. Cities like Lisbon and Oporto began to grow faster since the early twentieth century. However, it was in the 1970s and 1980s, with the increase of services' activities, that these cities expanded dramatically – often in an uncontrolled way, fuelled by market speculation – and leading to very poor-quality living environments.

One of the greatest benefits of the entrance of Portugal in the EU, in 1986, was the adoption of measures and regulations that have reinforced the efficient implementation of best practices of urban planning, namely, in terms of environmental impact. The Portuguese State devised three major policies to be followed in all interventions on the territory:

- The national strategy for social and economic development
- The national strategy for conservation of nature and biodiversity
- The national strategy for sustainable development

The interaction between these strategies resulted in a legal frame that includes all scales, from the regional scale of the regional territorial plans, which inform town planning, to the detail planning, which define what is allowed in each area with detail. There are also the very important coastal planning plans (POOC), which involve all coastal areas and discipline their occupation, preventing building in a strip of land along the coast. In addition to these, there are also the special territorial plans, to protect particularly sensitive natural areas, such as the Costa Alentejana or Ria Formosa in the Algarve, and limit seaside urban growth.

In terms of urban planning, and urban management, one can say that today, the State has all the necessary legal and effective means to prevent bad practices in terms of new construction and to correct existing situations resulting from the poor planning of the 1970s and 1980s. In terms of the latter case, much has been done, for example, on the eradication of illegal constructions by the seaside. The current refurbishment of the seaport of Sines, which will become one of the largest in Europe, also involved detailed environmental impact studies (Fig. 6.2).

Perhaps the greatest challenge for territorial management is to revert the growing trend of desertification of the interior, which is becoming dramatic, stopping the flow of people to the seaside areas of the country. A huge part of the EU and Portuguese legislation aims at valorising the interior, including the distribution of very significant subsidies to finance infrastructures such as highways, electricity, water and sewage systems, industry, schools, universities, hospitals, etc. But changing the trend of history tends to be a very slow process…. With this exception, one can say that the most difficult challenges in terms of the use of space that one can observe on the seaside are not so much related to the urban scale, but to the building scale – and in particular to tourism or "weekend house" buildings and resorts. Some bad and good examples will be shown next.

Fig. 6.2 *Left*: protected nature reserve in Costa Alentejana (*dark green*); *right*, *top*: beach at Costa Alentejana; *right*, *bottom*: project for the new port of Sines

6.2.2 Tourism

Most of the tourism activity of the world is centred on the seacoast. Portugal is no exception: being a country with modest resources, and away from the great industrial centres of Europe, tourism is a critical sector for the national economy, representing 10% of the GDP for decades.

Although tourism buildings and infrastructures occupy only a small parcel of the seaside, in comparison with the existing residential urban areas, it is a sector that tends to grow. This growth demands a great balance and good sense from the urban and building designer and most of all a solid legal framework – above all in regions with vocation for mass tourism.

A first study on tourism areas was carried out in 1964, resultant from reports of invited international experts, which identified three priority zones for tourism in Portugal: Lisbon area, the Algarve and the Island of Madeira. One of the first large-scale plans to order tourist occupation at urban level, just after the referred report, report was the DODI plan, for the Algarve. Other plans followed, such as the regional plan for tourism in the Algarve territory in 1992.

Fig. 6.3 *Above*: Praia da Rocha, Algarve. *Left*: before. *Right*: after the 1970s and 1980s. *Below*: São Martinho do Porto, *Right*: after the 1970s and 1980s

On the other hand, with the entrance on the EU, vast areas were incorporated in special plans, with environmental goals: such as the natural park of the Alentejo coast and the Ria Formosa Park in the Algarve. Today, all the Alentejo littoral is ruled by the PROTALI regulations – the regional plan for Alentejo.

The main challenge for the sustainable development of tourism lays not so much at the urban level, but most of all at building level. Improving architectural quality is essential. Buildings should be well integrated in their natural or urban contexts, especially in terms of scale and quality. Landscape architects should be consulted in this process. Bioclimatic design should be encouraged through legislation.

One can observe many examples of poor design throughout the seacoast, from isolated buildings that are dissonant with the surrounding context to whole areas of a village or city. The worse cases are generally from the1970s and 1980s when there was a "boom" of speculative construction – many small-scale, beautiful, fishing villages were turned into suburban-style concrete jungles, such as in Praia da Rocha, in the Algarve (Fig. 6.3).

However, side by side with terrible examples such as the presented above, one can also observe great design. Very good examples are the Hotel do Mar (1972), in Sesimbra, by Arch and Conceição Silva, or the very recent EDP museum, in Lisbon (2016) (Figs. 6.4 and 6.5).

In terms of resorts, Vilamoura, in the Algarve, was designed as a well-structured area, with all the necessary equipment: marina, hotels, sports fields, etc. It

Fig. 6.4 Hotel do Mar, Sesimbra, by Arch. Conceição Silva, 1973

Fig. 6.5 The Museum of Architecture, Art and Technology, in Lisbon, 2016 (*right* side)

remains a good achievement, together with Quinta do Lago, Balaia and Vale de Lobo, also in Algarve.

Vilamoura is a coastal resort and residential community, which started in 1966, and is located at the Algarve, offering a wide variety of apartments and villas for tourists and residents. In 1996, a new management strategy was applied to make Vilamoura more distinctive and better integrated into its natural surroundings. The plan included guidelines and incentives for construction companies, property owners, small business and hotels, minimization of waste, efficient use of resources, the protection of natural and cultural heritages and a contribution for raising environmental awareness and community activities. Strict control of construction and design was implemented, incorporating environmental considerations into the design project, reducing construction densities (50% or less than in previous phases), and occupation indexes; the green space was preserved over 90% of the land area; and traffic control mechanisms, cycle ways and car-free pedestrian pathways were introduced. At present, Vilamoura is extending its

Fig. 6.6 Vilamoura, Algarve

environmental management system (EMS) to infrastructure and residential development, which carries out regular internal environmental audits and independent external audits every 6 months. These interventions allowed the project to be awarded with the Green Globe Destination (since 2000), to become the first golf operator in Europe to have an EMS certified to the ISO 14001 norm (since 1998) and to have the first marina in Portugal with an EMS certified to the ISO 14001 (in 2002) and the European Golf Federation's "Committed to Green" program (recognized in 2002) (Fig. 6.6).

The next chapter will present the main long-term environmental challenge to coastal areas: the sea level rise, which urban design strategies can be used to mitigate its impact.

6.2.3 Environmental Impacts: Sea Level Rise

The previous chapters provided a brief overview on the evolution and present situation of the seaside urban design. It was shown that much has been done to solve the problems arising from the growing pressure of human occupation by the coast – namely, in terms of the legal frame that rules the density of occupation, need for infrastructures and nature reserve areas.

In terms of environmental protection, the influence of EU directives has led to a growing number of natural reserves, some of which occupying vast areas, such as in the Alentejo coast, where new building remains highly restricted. Pollution on the coast, on the beach resorts, also had a dramatic decrease in the last 20 years – most of the beaches along the coast have the "blue flag", signalling clean seawater. Of course one can – and should – always aim at higher environmental standards, on maintenance of what was achieved, and improving the natural space, but one can safely state that the legal and enforcing tools exist, and most of the work has been relatively well done.

However, there is a major long-term environmental threat, of a much greater scale than those previously referred: the sea level rise, caused by global warming (Fig. 6.7). Its effects have been monitored for years: in the last century, the sea level rose 20 cm on average on the coast; sea storms and floods have increased exponentially, leading to cliff erosion and serious damages on buildings and infrastructures, representing many millions of euros.

As it is well known, these are problems that depend primarily on the will and efficiency of international politics. However, Architects, urbanists, engineers, legislators and builders can present significant contribution in providing solutions to adapt or when possible mitigate the future reality.

The discussion on the problem of the sea level rise is relatively recent among architects and urban designers in Portugal. In terms of urban legislation, the previously referred POOC – the Coastal Areas Management Plan – is a very important tool, as it actually forbids any new construction closer than 500 m to the sea. Other urban design strategies can and should be used and new ideas brought about. The following case study is a good example of integrated urban design.

6.2.4 Case Study: Bioclimatic Urban Design in Northern Portugal

The littoral north coast of Portugal, mainly in the area between Gaia and Ovar, has suffered from the onrush of the sea over densely populated areas with permanent buildings that have been massively built since the middle of the last century, having succeeded in breaking the balance of the original and beaches, dunes and surrounding green areas, which operated as a thermal and environmental protection element (Fig. 6.8).

Such circumstances have breached the thermal and environmental balance that the previous urban settlements enjoyed over the centuries, both by the careful choice of places of implantation and by the low urban and population density that they have been able to preserve and understand.

Espinho, a coastal city located in this latitude, was a reference case because it grew and developed over the original beach still in the late nineteenth century as a typical summer activity city, having at the beginning of the twentieth century suffered with the advancement of the sea, destroying great part of the original city, imposing later on the redesign of the rest of the city located to the east part and serving as a warning to the same kind of urban settlement of others costal Atlantic cities, mainly about the risk of instability of the coastline and the impetuosity and strength of nature (Figs. 6.9 and 6.10).

This area is located on the north coast of Portugal, located at latitude 41° N, characterized by the topography in the form of an amphitheatre facing west, with smooth and progressive slopes and exposed to a large solar amplitude. The average annual solar radiation rate is 4.4 kWh/m^2, with a daily average of 1.7 kWh/m^2 in

Fig. 6.7 A simulation of the territory with a 100 m rise in sea level: the more densely populated areas in the territory – the coastal areas – would be completely flooded. The cities of Lisbon and Oporto would disappear almost completely…

January and 7.2 kWh/m^2 in July. The average air humidity varies between 60% in August and 80% in December, and the maximum monthly average temperature in Oporto varies between 13.1 ° C in January and 24.7 ° C in August, with January coldest and August the warmest, as Table 6.1.

Until the beginning of the twentieth century, this coastland was protected by dunes, vegetation and local pine trees, which controlled the violence of the winds coming from north and northwest, between May and September, reducing its speed and intensity.

Fig. 6.8 Aerial view of the Granja zone, near Oporto

Fig. 6.9 Images from the early 1900s – collapse of buildings over sea waves

Urban settlements were developed near the maritime coast, but they were sufficiently protected by the pine groves in a first line of protection until the end of the nineteenth century, with the appearance of small housing clusters belonging to fishermen. Later on, with the implementation of the railway line, the first summer-houses of wealthy families appeared in the beginning of the twentieth century.

Currently the coast is poorly protected by dunes, with a high lack of vegetation and native species suffer the action of maritime winds with high moisture content and salt of N and NW with some aggressiveness between the months of May and September. In the remaining months, the winds are felt to be S and SE, following the rains, according to the data obtained by the meteorological stations of the Oporto.

The sea currents are cooled by the very cold, deep-water currents (upwelling phenomenon) affecting the air masses coming from the west, causing the common fog during the bathing season (June to September), which lasts for almost all the day.

The climatic characteristics clearly differ from the inland waterfront (at least 1 km away), partly due to the barriers of existing forests, attenuating the NW winds and fog in the summer months, creating a milder climate throughout the year.

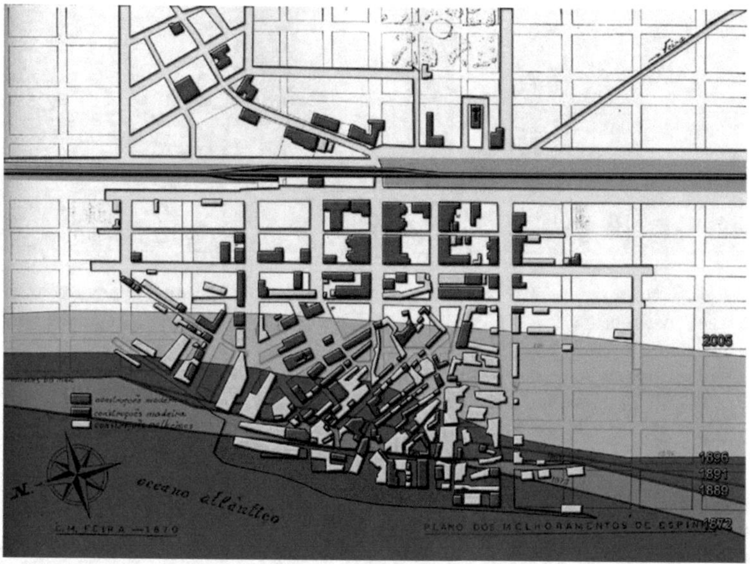

Fig. 6.10 Map of Espinho – progressive costal line movement since 1872 until 2005

Table 6.1 Temperature in Porto

Temperature in °C	Jan.	Feb.	Mar.	April	May	June	July	Aug.	Sep.	Oct.	Nov.	Dec.
Month mean value of minimum temperature	4.9	5.5	7.1	8.7	11	13.6	14.9	14.7	13.9	11.1	7.8	5.7
Month mean value of maximum temperature	13.1	13.9	15.7	17.5	19.4	22.3	24.3	24.7	23.6	20.4	16.1	13.6

The configuration of the urban settlements that were less planned and regulated has undergone an excessive densification of the soil, particularly in the seafront, reflecting a greater transfiguration in the environmental, urban, quality of use of public space and general precarious construction quality along the coast, due to the lack of implementation and experience of urban planning instruments.

Diagnosis of common negative aspects:

- Construction of buildings as concrete corridors acting as wind tunnels, improving worsen climatic elements and providing the misaligned shade of contiguous and parallel streets and buildings, without any concern for adequate spacing in order to avoid shadowing contiguous buildings (Fig. 6.11).
- Uncontrolled penetration of sea winds, with high levels of moisture and salinity, leading therefore to more wear and tear in the constructions themselves.

Fig. 6.11 *Left*, construction of concrete corridors as wind tunnels; *centre*, absence of trees in the streets; *right*, inadequate urban framing

Fig. 6.12 Uncontrolled penetration of sea wind

- Destruction of old pine groves, which, together with the absence of vegetation along the coast, does not allow the reduction of wind in the area (Fig. 6.12).

- The absence of strategies in the design of new blocks denoting the wrong options for both the size and the orientation and shape of the buildings, privileging only economic factors and ease of execution, adopting guidelines favourable to the propagation of sea winds in the summer and south and southeast winds in the winter.
- Existence of streets unprotected from the wind because of the absence of trees.
- Absence of protection galleries in the main pedestrian courses.
- Compromising of potential bioclimatic strategies adopted in architecture, because of an inadequate urban framing, originating bad thermal conditions in the building(s).
- The combination of the above-mentioned factors affects, in a negative and probably irremediable way, the quality of the urban territory we are dealing with.

6.2.4.1 *Definition of Strategies and Application of Solutions for Urban Bioclimatic Design*

The territory under study has thermal constraints that are unfavourable for buildings and in particular for its users, both in the built environment and in the public outdoor space, and it is necessary to optimize the local conditions for better energy efficiency, as well as ensure minimum levels of quality of use of the public space.

For this it is necessary to distribute and orient properly the buildings, in order to avoid shadowing to the surrounding constructions and to promote an optimization of direct solar radiation; for that, we should consider a 25° angle for a latitude of about 41° N, as the maximum angle that will define the relationship between the spacing and the height of the building immediately towards south and therefore causing possible shadowing measured in the small day of the year – the winter solstice at solar noon.

Other climatic and environmental factors to be considered in the definition of bioclimatic strategies for the territory under study:

- Solar geometry (orientation)
- Geomorphology (slope relief, slope exposure, altitude)
- Urban space (dimension, shape and urban design of the territory)
- Influence of winds (wind flows, aeration, layout/orientation of streets and public spaces, cold areas)
- Vegetation and distribution of green zones (buffer zones, thermal regulator, oxygen regeneration, vegetation shading)
- Hydrological cycle (water cycle, water balance)
- Levels of comfort in the urban environment

Possible strategies to be applied at the level of urban planning as possible solutions to ensure a better urban quality under bioclimatic standards:

Construction

- Definition of maximum heights to be built appropriately to the profile of the street, orientation and topography, in order to limit the shading areas between buildings and their roads.
- When the topography imposes slopes higher or lower than 7°, this inclination should be taken into account for the calculation of the relation between spacing and distance from buildings.
- Definition of maximum density of soil occupation, articulating with the definitions of volumetry and with defined road profiles.
- Definition of distances to the street according to the orientation of the streets (axis N-S and E-W with different measures of distance to the construction), with greater distance in E-W streets than in the case of streets N-S.(Fig. 6.13)
- Distance from opposite facades of the buildings on the street should be considered according to the orientation of the streets (where the NS and EW axes must have different measures of distance from the opposites facades), with greater distance in EW streets than in the case of streets meaning NS.
- Use of covered traffic galleries in trade and service areas.

Roadways

- Proposal of new streets taking into consideration: orientation, topography, dimension and density of construction attached to them and implantation of new trees (Figs. 6.14 and 6.15)
- Draw roads towards E-W with greater width to ensure complete solar shading without front of the buildings.
- Avoid the design of urban canyon too long in a straight line, opting for traces with some variation, should be preferentially undulating, curvilinear and never rectilinear, functioning as a decelerator of winds (Figs. 6.15 and 6.17).

Fig. 6.13 Orientation of the construction volumes towards south

Fig. 6.14 Streets with less than four intersections

- At the intersection of E-W streets and N-S streets, you should always add evergreen trees to absorb and reduce wind aisles (Fig. 6.16).
- Decentralize and reduce intersections from four to three streets, acting as attenuating for winds.
- Main traffic flow routes should be oriented in the N-S direction, thus ensuring E-W streets a less direct (more undulating) route, more residential and local.

Fig. 6.15 Streets and paths too long in a straight line

Fig. 6.16 Streets with evergreen trees

Vegetation

- Preserve pockets of local vegetation, promoting green protection parks, thus allowing to control humidity levels, to guarantee minimum levels of permeability of the soil and to guarantee the protection of water lines, allowing only low-density constructions to be built in direct envelopes (Fig. 6.18).
- Ensure street-side solutions with trees.
- Design of spaces for public facilities with green zones, low density and occupation of the ground, such as outdoor playground, golf, tennis, volleyball, maintenance circuits, urban parks, etc. Allowing to preserve green bags in the territory

- To create barriers to protect the action of sea winds by inserting permanent leaf trees, preferably soft pine, by their large and small canopy, releasing the deciduous trees to the buildings, with seasonal characteristics important for their thermal optimization (Figs. 6.18 and 6.19)

Fig. 6.17 Streets with less than four intersections

Fig. 6.18 Streets with evergreen trees

Fig. 6.19 Public equipment – urban parks

6.3 Building Design

6.3.1 Sustainable Architecture

Sustainable architecture aims at producing buildings that are adapted to local social-economic, cultural and environmental contexts, having in mind the consequences to future generations. Within this frame, the top priority must be to minimize energy consumption in buildings (both in terms of maintenance and embodied energy), through the use of passive design strategies, i.e. reducing the use of energy-consuming equipment like HVAC or artificial lighting, through a wise adaptation of the building to the local climatic context.

Although, on average, the energy consumed by buildings accounts for only 25% of the national total, this figure rises up to nearly 40% in the larger cities, where most of the population lives. The foreseen tendency, if no effective measures are taken, is for the national average to keep increasing in the next few years and approach the EU average of 40%. Considering that more than 80% of the energy source used in Portugal is (imported) oil, this means that professionals of the building sector, such as architects, engineers or builders, have a serious responsibility in terms of their contribution for inverting this tendency and promoting a more sustainable development.

In Portugal, as in most European countries, research on energy efficiency in buildings emerged in the 1970s and was pioneered mostly by civil or mechanical engineers. At the time, research was primarily motivated by economic concerns. However, during the past two decades, with the growing awareness of the environmental problems caused by energy consumption, the emphasis of scientific discourse (rightly) shifted from economic to environmental issues. It was mainly during the late 1990s that architects became aware of the importance of this subject.

Many of the passive design strategies, such as natural ventilation, solar orientation, the use of thermal inertia, shading, etc. are basically an adaptation of techniques used in the past, resultant from centuries of accumulated empirical knowledge, to contemporary requirements. Unfortunately, this knowledge was progressively abandoned from architectural practice and teaching since the implantation of the modern movement in the first decades of the twentieth century: as energy was cheap and the environmental problems we share today were not yet known. As a consequence, we inherited an "international style" architecture, dissociated from climatic and social-economic contexts, with extensive use of artificial lighting and climatization, seen in many of buildings across Europe.

Portugal has a privileged climate, relatively tempered, in which the use of artificial climatization is unjustified in most situations, if the architectural conception is adequate (Fig. 6.20).

In terms of the use of the passive design strategies, it may vary, reflecting the climatic diversity of the country, from temperate Atlantic in the coast to Mediterranean and continental in the interior areas. This can be clearly observed in the large variety of vernacular architecture solutions found throughout the country. However, we can generally identify a significant number of common techniques and principles applicable to the various regions of the country, as in other southern European regions, such as night ventilation associated with thermal inertia (for cooling in summer), proper solar orientation, adequate insulation and dimensioning of the glazing areas (for solar heating in winter and protection from overheating in summer), control of the building depth, adequate shading, etc.

The same can be said in terms of artificial lighting: like our southern European partners, we enjoy of excellent levels of daylighting, a fact that is in flagrant contradiction with the (acritical) "importation" of "international style glass towers", observed in many of our cities – causing serious overheating in summer and unnecessary need for heating in winter.

Active systems, such as solar thermal and photovoltaic, or low-energy hybrid systems, such as evaporative coolers, also offer a great potential for reducing energy consumption from non-renewable sources. However, despite the technology being available and widely tested, its market still has much to grow, especially when compared with other EU countries with much lower levels of solar radiation such as Germany or Denmark. Although these systems should always be regarded as a complement to architecture to improve building performance (and not a remedy for poor design), their integration in the building's design is an interesting challenge to architects.

In terms of sustainable development, the situation presently found in Portugal offers good opportunities in two critical areas: building refurbishment and the revision of comfort criteria.

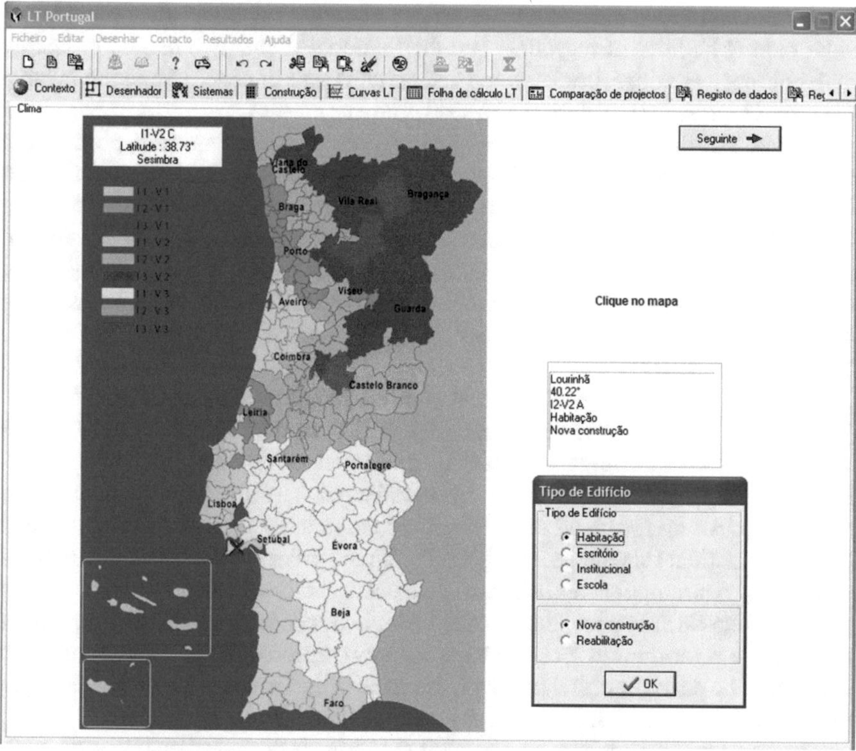

Fig. 6.20 The LT Portugal software, developed by the Higher Technical Institute and the University of Cambridge, is a user-friendly program to calculate energy consumption based on (a) local climate, (b) building geometry, (c) insulation and glazing ratio, (d) thermal inertia and (e) shading devices. It can be used at the early stages of design, to compare different solutions – both in terms of new design and refurbishment. The first step is to choose the location of the building – and its climate. The different colours of the map result from a combination of three winter and summer zones (severe, moderate, mild). Notice that the seaside coastal area presents the milder, most temperate, climate combinations

In Portugal, the great majority of existing buildings are (still) naturally ventilated, and a large percentage of these were built before the introduction of mechanical means, i.e. are, in essence, “selective” buildings.

The practice of sustainable architecture is an essential factor for a sustainable development. The conception process must integrate not only bioclimatic strategies but also questions regarding the social-economic impact of the building throughout the different phases of its existence. For example, the choice of building materials has implications not only in terms of energy consumption (embodied energy) but also on local economy. Preference should be given to local material and human resources, thus promoting regional development and avoiding the environmental impact of transports. It is also essential to consider the lifetime of the building, which should be prolonged to the limit, avoiding the energy and economic costs of demolition and new construction. The choice of construction materials should consider the possibilities of recycling, and their embodied energy – using low-energy materials such as adobe, bricks, wood and certain types of concrete and minimising the use of high-energy materials like steel, glass or plastics.

Despite the fast-growing sales of packaged air conditioning units observed in recent years, there is still time and opportunity to put a stop to this tendency, through the promotion of sustainable rehabilitation measures, when necessary. Existing knowledge on environmental design strategies is today able to provide solutions to most situations, even to critical ones like deep-plan "glass towers". In the latter cases, for instance, it is also possible, if economically unavoidable, to establish compromises, through the application of "mixed-mode" solutions or the use of "solar HVAC" systems.

The revision of the present thermal comfort standards is also a major priority. International standards such as ASHRAE's (USA) or the EU ISO 7730 (that provides the basis for Portuguese standards) are contested by the results of fieldwork carried out in the past three decades, by scientists such as Humphreys, Nicol, Baker or De Dear. A large segment of the scientific community proposes the elimination of the conventional standards, considered to be too inflexible, implicating unnecessary energy consumption, and often causing occupant dissatisfaction, or even health problems like SBS. Instead, the so-called adaptive school proposes the adoption of the adaptive algorithms to form the basis for new comfort criteria – which considers comfort to be strongly correlated with the local climatic context, hence offering a flexibility that provides support to the practice of bioclimatic architecture.

The event that probably most contributed to trigger this awareness was the realization of the Expo, where an emphasis was put on energy efficiency, both in terms of building and urban design (Figs. 6.1 and 6.8).

Since the late 1990s, an increasing number of architects have become interested in sustainable, low-energy design, and as a result one can find some very interesting built examples of contemporary "environmentally conscious" architecture in the country, among which are the "Green Towers", by Arch. Livia Tirone, and the "Solar XXI Building", designed to lodge the Building Research Institute (INETI).

The "Green Towers", in Lisbon, achieved a reduction of over 40% in energy consumption, compared to a conventional building, and the occupants' level of satisfaction is greater than 90%. The main passive strategies used are external insulation, adequate orientation, glazing ratio and shading and the use of Trombe walls for heating in winter (incorporated in the south facades).

The INETI building, finished in 2005, aims at achieving "zero GHG emissions", with the aid of active solar systems – it serves also as an experimental laboratory to study the efficiency of other low-energy strategies such as ground and evaporative cooling systems. The INETI building is one of the rare examples of building-integrated photovoltaic (BIPV). There is an excellent potential BIPV in Portugal. The mean value of insulation decreases from north to south with a minimum of about 1600 h and 2200 h in the NW (Alto Minho) and the minimum of 2600 h to 3300 h in the east part of Alentejo, in the south coast and in the Lisbon region. Despite this potential, there are still very few examples of BIPV in Portugal, mostly due to its relatively high costs. Recent studies demonstrate that Portugal could decrease its greenhouse gas emission through BIPV and up to 140 million euros if efficient legal instruments were created in relation to efficient construction and solar integration.

The EU has also played an important role in promoting energy-conscious building design in the last decade, through design competitions and research programs such as Save. Building and services' regulations have again been revised in 2006, with an emphasis on energy-saving issues, much improving the previous standards. After January 2008, all new building projects must have had an energy efficiency certificate, attributed by a specialist, as a mandatory factor for approval.

In terms of the use of renewable energy systems, new policies were also implemented in 2007, providing attractive tax benefits to users, particularly in what concerns solar (thermal) and photovoltaic panels.

Building promoters are starting to search for a "green label", and the need for refurbishment – as opposed to new construction – is taking roots. These facts also show a raising awareness of environmental issues by the market, presenting a good opportunity for the implementation of sustainable urban and building design.

Universities have also been active in the field; a growing importance has been given to the area of sustainable architecture and urbanism, both in terms of teaching and research. Design tools such as the LT Method or Insight 360° (REVIT) are being used by a growing number of students and perfected by research.

6.3.2 Comfort Criteria

Our quotidian cycle encompasses a diversity of activities, which may lead to fatigue and recovery. To counterbalance the mental and physical stress due to activities occurred during the day, it is vital that mind and body recuperate through rest and recreation. Unfortunately, unfavourable climatic conditions may impede recovery. This break in the cycle results in discomfort and causes mind and body to be fatigued. Furthermore, these weariness and stress lead to loss in productivity, loss in efficiency and health breakdown. Therefore, it is necessary to be aware of the importance of the role and effect of climate on man.

It is the duty of the architect to create the best possible outdoor microclimate and consequently an improved indoor space. The quality of a building will be given by the physical as well as emotional conditions of the occupants. The optimum of complete comfort which means the total physical and mental well-being is a goal to be sought by architects.

As it is difficult to analyse the emotional aspects, comfort criteria usually emphasize thermal comfort and physiological responses to different climatic conditions. Human response to the thermal environment is dependent on air temperature, humidity, radiation and air movement.

Power and material requirements of the body are supplied through metabolism, which converts ingested food into useful energy and matter. From the total energy produced in the body, only 20% is utilized, while 80% is extra heat, which must be dissipated to the environment. Therefore heat is continuously being produced by the body. Not only the surplus heat from the body but also any heat gain from the environment, for example, solar radiation, must all be dissipated in the surroundings. This is necessary to maintain body temperature around a constant and balanced 37 °C. In this give and take relationship, if human responses to the environment are to be predicted, it is vital to consider all thermal effects and produce the best indoor and especially outdoor microclimate.

Microclimate and building thermal performance can be significantly affected, and energy cooling loads reduced by the modification of air temperature, solar heat gain, long-wave heat gain and heat loss by convection as well as humidity levels may also be modified, through the careful use and interaction of the natural environment with the built environment.

6.3.3 Bioclimatic Strategies

The design integration of both these environments leads to the concept of bioclimatic architecture or sustainable architecture. A bioclimatic profile offers a set of information, conditioners and determinants to be taken into consideration in architecture and urban design to find adequate solutions. A major concept about bioclimatic architecture is one of saving energy. For this it considers three basic fundaments of architecture: the place, the culture and the history. This may be translated into the concepts of *locus* and *situs*, or *context* and *surroundings*, providing information on the climate, microclimate, materials, landscape and building forms. The best interpretation of all this information, through the integrating relation of aspects such as thermal, luminous and acoustic, ventilation and terrain, based on a certain architecture program, a certain landscape, culture and well-being notions, denotes a qualitative habitat, a habitat that preserves its surroundings, opens to environmental comfort and promotes prosperity, enhancing architecture, city, nature and man. It evokes the *genius loci*!

6.3.4 Materials in Seaside Areas

Designing and building on the coast requires extra care from the designers that specifies the materials. When reacting with the sea air, they may present atypical behaviour. It is necessary to know how to use the available solutions correctly and the characteristics of each raw material, so that the sea air does not interfere with its performance.

Wood may be the raw material for the window frames, but in this case, the hinges must be made of stainless steel to prevent corrosion caused by the sea air. If they are painted, the wooden frames should be previously sealed or use a primer for protection. Aluminium windows and doors are corrosion resistant, but the ideal finish is electrostatic or anodizing paint, which guarantees low degree of dirt adhesion.

To achieve the perfect seal, the rubbers used around the windows need to be replaced, on average, every 3 years, depending on wear. On the other hand, PVC frames are immune to corrosion, having good sealing and thermal performance. Regardless of the material, the frames must receive lubrication of the joints at least once a year.

Wood can be used in coastal regions, but requires extra care with maintenance and the application of varnish at least once a year. Wood solutions have to be well analysed due to the fact that the sun and rain exposure accelerates its deterioration. To avoid problems, the appropriate species must be specified, and the material must undergo treatment with the application of wood seals that create a protective film.

On the other hand, masonry has does not have much interference from the sea, as it is usually made with ceramic or concrete blocks. However, because coastal towns are wet, if finishings are not well made, the tendency is for moisture to seep into the walls. The finishes of simple mortar, made with cement and sand, also suffer from the effects of moisture, so an alternative is the ceramic or stone finishings, as they do not suffer with the sea air. Paints nevertheless fade when exposed to the sea and sun, especially stronger colours that fade even faster.

Some types of paint may last longer. Acrylic paints applied to external walls last for about 3 years, while the rubberized and elastic, elastomeric bases have a durability of 5–6 years. In indoor environments, the best option is semigloss acrylics, which visually soften the action of the sea and have good resistance.

In structures, iron is the element most attacked by the sea. To avoid problems, the fittings must be protected by concrete. The specific calculation standards for concrete overlays in beach buildings determine more volume than would be necessary away from the coast. Other metal structures also require care, and the steel must receive paints against corrosion.

Ceramic tiles need to be waterproofed, to lessen the crystallization effect of the sand salts, which generates crumbling of the finish. Concrete tiles should have a layer finish of acrylic lacquer, which ensure clean roof for longer.

Furthermore, in buildings with flat roofs, it is possible to use asbestos cement tiles without asbestos or waterproof slabs. Descending from the roof to the floor, the most suitable are the ceramic tiles with matte finish. Stained tones may disguise the presence of sand.

Steel and concrete were joined to create reinforced concrete, one of the most used materials in construction, making it possible to raise buildings, monuments and constructions, which challenge gravity and impose beauty. Much of the success of this combination is justified by the fact that the frames used for reinforcement do not in theory require corrosion treatments. This is due to the high alkalinity of the concrete, which favours the formation of a stable oxide film that pacifies the steel and prevents the progression of corrosion.

Nevertheless, this combination is not always harmonious. Pathologies can occur when there are vulnerabilities in the system, mainly resulting from failures during execution and in the concrete specification, which should be adequate to the exposure situation. In this sense, one of the most aggressive agents to concrete structures is the marine atmosphere, where sodium and magnesium chlorides are suspended in the air. On the coast, salts are removed from the sea by waves and transported by air. They can travel long distances and settle on concrete in the form of water droplets. The biggest problem is chloride ions, which are very small and with high mobility in the interior of the concrete.

The corrosion rate in a marine environment can be 30–40 times higher than in a rural atmosphere. Damage can be even more severe when the structure is directly subject to sea level variation (splash zone effect). This is because, besides the attack of chlorides being more intense, in these cases, there is formation of secondary ettringite, material of expansive character, by the reaction of cement aluminates and sulphates of seawater. The crystallization pressure of this component is very large with an expansion of more than 300%, so when this pressure reaches the tensile strength of the concrete, cracks occur and the concrete deterioration process intensifies. In the face of sea air exposure, all types of concrete structure can suffer attacks of chlorides, sulphates and other aggressive agents. However, those facing the prevailing winds are more susceptible, as are sites with a large surface area in relation to volume, pillars and beams, especially. The problems increases in risk and seriousness, as the structural element gets slimmer.

Other factors may also contribute in triggering pathologies. Humid places with a higher risk of condensation, such as bathrooms, kitchens and laundry areas, tend to exhibit symptoms of corrosion in a quicker and more intense pace than in dry environments. In the same way, places with low ventilation are more subject to corrosion, as they may present mould and fungi that release acidic organic products into their metabolism.

Avoiding or minimizing contamination is, fundamentally, the basic and necessary principle of concrete structures located in coastal areas. Design solutions, such as positioning and location of the building in relation to the seafront, and specification of materials that are more resistant to environmental aggressions, are of utmost importance. In addition, durability can be ensured by the use of more impermeable concrete, the use of low water-cement ratio and the use of additives such as active silica, for example, to reduce porosity. The use of blast furnace, pozzolanic or sulphate-resistant cements, which perform better with regard to durability, is recommended.

Executive procedures regarding the transportation, launching, densification and curing of concrete as well as treatment of concrete joints and the deformation plan of the structure should also be closely observed. Spacers and fasteners must be executed to perfection. When they are poorly made, these elements do not fulfil their function of guaranteeing the original position of the armature during the concreting process.

In specific cases, when it is not possible to obtain the appropriate minimum covering or when there is no way to prevent the access of aggressive agents in the structure, it is possible to use resources such as galvanization of the armature, chemical inhibitors, which act on the metallic surface or concrete surface impregnation with impermeable coatings. Equally important are specific measures of maintenance, use, monitoring and inspection during concreting procedures and structure activity.

Regardless of the solution, it is fundamental to study all the conditions and the characteristics of the work and choose among the most appropriate procedures. If the structure is apparent concrete, for example, it is important to use highly effective varnishes or protective systems against the penetration of chlorides, such as a double compound system consisting of a silane-siloxane primer and a topcoat consisting of a noble acrylic or by methyl methacrylate.

However, the main precaution refers to the concrete covering, which should physically protect the armature and provide a medium with high pH for the passivation of the steel. Among the main innovations of the standard is precisely the requirement of larger coats and smaller values for the water-cement ratio in structures exposed to aggressive agents, such as the marine atmosphere. Concern with the durability of the concrete structures raises the quality of the structures exposed to aggressive environments.

Chlorides are the most aggressive agents to concrete structures. They can be present in the atmosphere in many ways besides seawater. In cold countries, for example, salt is used as a defoaming agent. At the other extreme, fire can also contribute to the suspension of chlorine if the fire strikes PVC pipes. In more polluted regions, chlorides may be present in the soil and reach buried structures of foundations. In the water reservoirs, chlorine is added for disinfection and reaches the inner surface of the concrete. Finally, chlorides can be found in materials that make up concrete, such as water or contaminated sand.

Structure pathologies are common in buildings close to the seaside. The most typical phenomenon of concrete structures exposed to the marine atmosphere is armour corrosion. It is an electrochemical process in which there is an anode and a cathode. The water present in the concrete serves as an electrolyte. Thus, any potential difference between points can generate a current, initiating corrosion. Generally the problem manifests itself by the reduction of the section of armature and cracking of the concrete, but, eventually, reddish stains produced by the oxides of iron can appear. The causes are varied, among which the insufficiency of the cover of the armature or poor quality of the concrete and presence of chlorides.

Another pathology is from the stresses caused by increased corrosion, where other problems may arise. First cracks occur because corrosion products occupy greater space than the original steel. Then, other pathologies can also affect the structure, such as disaggregations.

In environments with high level of aggressiveness, exposed to sea air, fissures and breakdowns are common. The opening of cracks in the surface of reinforced concrete structures should not exceed 0.3 mm. In places in contact with tidal droplets, the maximum acceptable aperture is 0.2 mm.

Associated with fissures is disaggregation, which is the physical separation of concrete slabs. The main consequence is the loss of the capacity of resistance to the requested efforts.

Another process common in seaside buildings is carbonation. Over time, high alkalinity on the exposed surfaces of concrete structures can be reduced, which is mainly due to the action of acid gases, such as CO_2, SO_2 and H_2S, found in the atmosphere. It is a process that occurs slowly. The precipitation pH of $CaCO_3$ is about 9.4 (at room temperature), which alters the chemical stability conditions of the steel passivating film. It is, therefore, a phenomenon linked to the permeability to the gases and, therefore, demands care regarding the concrete composition.

Lastly, expansion is one more usual pathology process, which may be found in seaside buildings. In the cement manufacturing, the used plaster reacts with part of the tricalcium aluminate forming ettringite. Another part of the aluminate is free to react if it subsequently encounters sulphates, present in aggregates and in seawater, with which the concrete will come in contact, producing more ettringite. As this occurs in a phase where the concrete is already hardened, pathological effects will appear in the form of cracks, fissures and subsequent disintegration of the concrete.

6.3.4.1 Final Considerations

Technology nowadays conflicts and tries to conquer nature. It works independently and in reclusion. It looks upon nature for questions, not answers. Together they can produce better results. Environmental issues are usually situated on a global scale. Discussions include global warming, damage to the ozone layer, overpopulation and scarcity of energy, resources and food. Nevertheless, damage is also caused on a smaller and daily scale, in a microenvironment with its own microclimate.

We think of fantastic and megalomaniac solutions instead of being practical and use common sense. We want to colonize the moon before conquering our own problems on earth or even our surroundings. Being instinctive is not being naive. Being simplistic is not being underrated. It means being rational, natural, intelligent and sustainable. It is giving a step at a time and not one bigger than the legs can reach.

The historical precedence illustrates the challenge of survival of mankind in a healthy, sustainable environment. Alvin Toffler (1980) in his book *The Third Wave* describes the third wave of progress and development in technological advances. Knowledge is the wave of the moment, not only scientific knowledge but also the rescue of common sense in the form of popular knowledge. The speed at which modern life is lived is much dictated by duties that reason has faded into the background. As part of the animal kingdom, we naturally pollute, destroy and modify. On the other hand, counterbalancing our actions, the plant kingdom recycles our resources, reacts towards the damage and provides us with the bare essentials needed to live, food, air and water. The intelligent and reasonable balance between both parts will ensure a healthy environment and consequently our future. We cannot attempt to resolve global issues without tackling local problems. We cannot attain a sustainable community without providing basic need to the lower, yet frequently majority, class. We cannot have healthy environments without comfortable microclimates.

Environmental design contributes to diversity and addresses formal and spatial issues that are not universal and cannot be globalized, but, on the contrary, are related to a determined context and microclimate. This incentive to diversity enriches spaces and lives. It admits architecture to be rich and interesting instead of vague, unsubstantiated and indistinct. Contextual architecture permits uniqueness rather than superficiality.

Environmentally friendly houses do not have to look awkward like prototypes or scientific models. They do not have to be ugly to be efficient. They can be simple and cheap or complex and expensive. It can be a humble house in a popular urban settlement or a last-generation high-tech "intelligent" building. They can be of any form desired as long as they incorporate and benefit from their surroundings. An agreeable microclimate will provide the user with an agreeable space. Comfortable, pleasant and healthy outdoors will certainly also affect and reflect indoors.

6.3.5 *Case Study: Espinho*

Based on the same city studied – Espinho, Portugal, and on the perspective for presenting guidance strategies, organization and proportions of the building in the demand for better bioclimatic and sustainable configuration, we developed a study on a proposal of urban design and architecture presented by the international team that won the international competition for the arrangement of the upper surface of the previous railway that crosses Espinho very close to the sea (Figs. 6.21 and 6.22).

Fig. 6.21 Public equipment – urban park

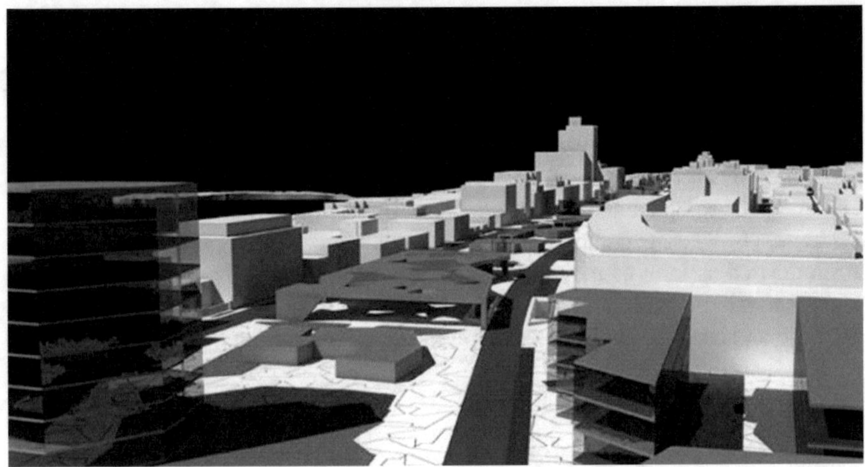

Fig. 6.22 Conceptual image of winning proposal for Espinho

The winning proposal of the competition by the architects Rui Lacerda, Francisco Mangado and João Álvaro Rocha (hereinafter referred to as the original scenario) deserved our attention in order to evaluate them from the point of view of solar access in summer and winter the performances of the 23 constituent volumes of the proposal.

6.3.5.1 Original Scenery

The original scenario consists of 23 residential buildings, identified by letters as shown below:

The volumes presented for the western zone of the railway, we mean to present a set of bioclimatic concerns that we could hardly overcome without prejudicing the solutions of integration and urban design recommended. The volume A (comb-shaped) as shown in Fig. 6.23 below shows a distancing between volumes of the correctly sized comb, taking during the winter solstice the necessary clearance to ensure solar access to all floors facing south. In the case of summer, since the facades with the largest exhibition area are oriented to the south and north, these are more easily protected than the facing facades faced to the west side. We understand, therefore, that this building presents a solution closer to the ideal one for a construction area and maximum height allowed in this plot. The main volume, where all the smaller volumes are interconnected, may guarantee a reasonably protected facade of the northwest winds and also achieve own protection by the smaller volumes of east-west direction for summer shading purpose.

The volumes with more irregular shapes at east of the volume A (comb) presented with higher height have a correctly articulated location between them, allowing to guarantee to all of them a balanced solar access, as well as the distance

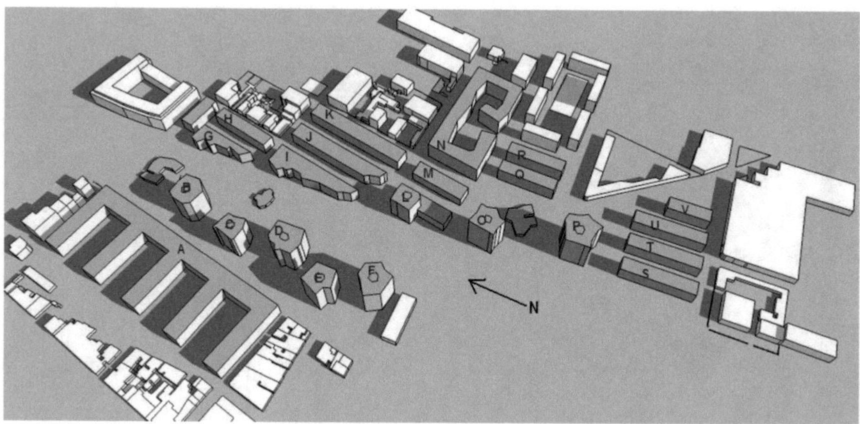

Fig. 6.23 Original scenario at 12 noon on 21 December with identification of volumes

between these westbound volumes from the railway to the volumes on the east side of the railway is understood to be adequate since, under tests at the winter solstice, only at the end of the afternoon those volumes located on the east side of railway are affected by the shadowing of the volumes referred above.

Thus, we consider that all the solutions recommended for the area at west side of the railway are well designed according to bioclimatic principles.

As far as the east side of the railroad is concerned, the presented solutions are, in our opinion, fallible for both summer and winter and have received new proposals for urban organisation, guaranteeing the same maximum height, and also have the same building area defined by the authors in the original plan.

6.3.5.2 General Strategies Adopted for Design Proposals

In the pursuit for better thermal performance, both in summer and winter, we developed two proposals (1 and 2) with a set of bioclimatic principles based on bibliographical references of several authors.

The most important strategies for the design of models 1 and 2 are:

- Orientation of volumes with their greater exposure of facades facing south
- Providing adequate spacing between volumes to ensure solar access in the winter
- Reducing the form factor of the volumes while maintaining the same building area
- Protecting facades facing the west, specifically in the cooling station

6.3.5.3 Analysed Models

We analysed three models: the original model and two proposals.

With proposal 1 and having as the starting point the design of the original scenario, we redefined the orientation of most of the volumes, privileging the

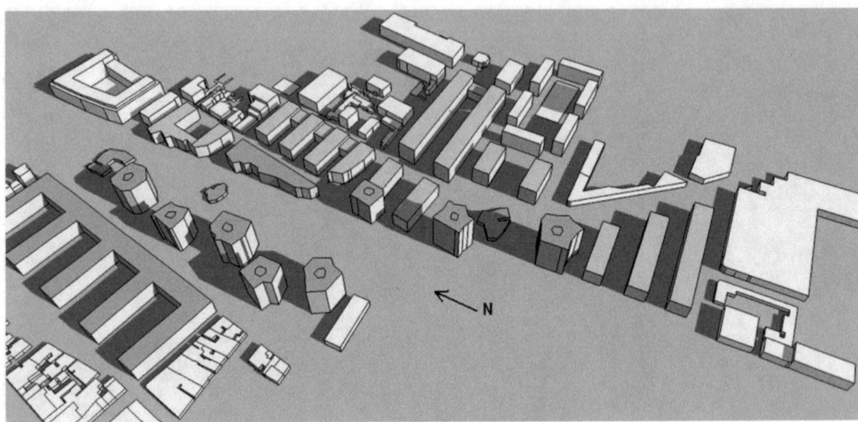

Fig. 6.24 Proposal 1 (*yellow* buildings) at 12 noon on 21 December

orientation of volumes with their greater extension facing south, as opposed to the volumes of the original scenario, mostly oriented to the west and east. For a formal urban design issue adopted by its authors, we decided to preserve the design of a significant set of buildings, which we consider to be very relevant for the identity of the winning project (Fig. 6.24).

Of the final solutions presented as better than the original solution, proposal 1 is precisely that which, gathering the knowledge obtained in the studies of the urban matrix, lists a series of principles known and tested previously, and that, in a more conscious way, we understood as the urban solutions capable of obtaining better performances in the solar access in the winter and protection to solar radiation in the summer, when compared with the original proposal of the authors.

We tried to reduce the volume solutions with excessive areas of solar exposure at west and east sides, or those that did not offer any protection coming from contiguous volumes, choosing, alternatively, solutions of orientation of the facades with greater extension facing south and north. Only in the volumes that we understood that could compromise the idealized urban framing solutions, closing streets or changing buildings shapes, did we choose to keep the original solution or trying detracting from it as little as possible (Fig. 6.25).

The proposal 2 arises from the need to, in order to overcome the performance obtained with proposal 1, add the higher value of integrating concepts of qualification of the urban space, guaranteeing zones of leisure and well-being for the users. In addition to climate concerns and energy benefits, we seek qualified urban solutions where the street would cease to be just a wind corridor, creating rhythms that could attenuate erosive effects of the wind and concomitantly boost solar gains by the configuration of the design of the patios.

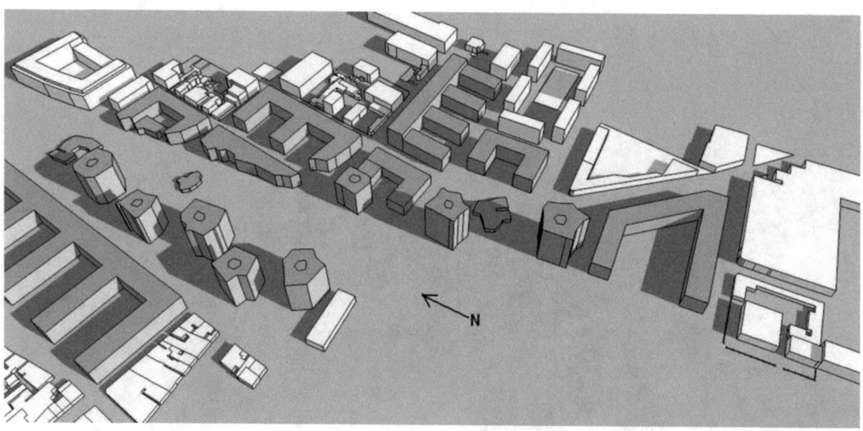

Fig. 6.25 Proposal 2 (*blue* buildings) at 12 noon on 21 December

6.3.5.4 Definition of Design and Analysis Criteria

For the evaluation of different solutions (3) to the units (1 to 6) chose to focus models with smaller form factor.

The global solar radiation values are presented in kWh/m2, weighted average of all volumes for winter period (15 January and 21 December) and summer period (1 August and 21 June), and in kWh, total radiation value of all volumes for the same periods of the year.

6.3.5.5 General Evaluation

As can be seen in Figs. 6.26, 6.27 and 6.28, the buildings located in the western zone of the railway show that the weighted averages of total solar radiation on the typical day of the heating season (15 January) very close together, with values in the range of 1,80 kWh/m^2, compared to the area east of the railway where the average results are on the range of 1.70 kWh/m^2. But the most relevant is the fact that in the area east of the railroad, there are differences between volumes in the range of 0.55 kWh/m^2. It is justifiable to present new solutions of volume design with better results in solar access in both analysed periods.

In the typical day of the cooling season, the volumes in the zone west of the railroad show values very similar to each other, in the order of 4.40 kWh/m^2, which could be attenuated with other solutions of shape of the buildings; however we would not be able to do it without reducing the recommended building area or maximum limits defined for this area. Thus, we decided to consider as an effective solution by the implantation geometry appropriate to the specific characteristics of the direct envelope and the maximum allowable boundary (Figs. 6.29, 6.30 and 6.31).

In the zone east of the railroad, the values obtained between buildings vary in the order of 3.43–4.72 kWh/m^2 with an average of around 4.15 kWh/m^2.

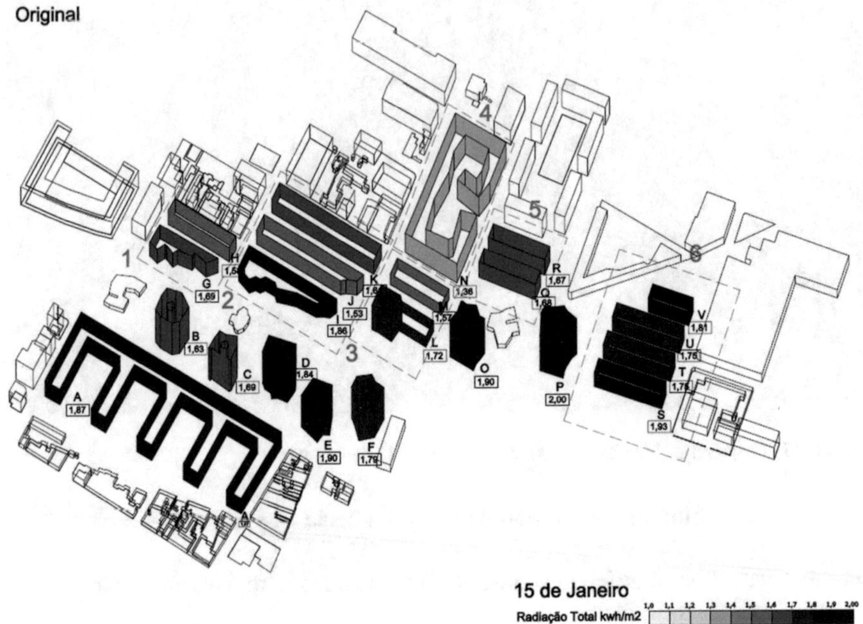

Fig. 6.26 Overall analysis of the original scenario for the typical day of the heating season

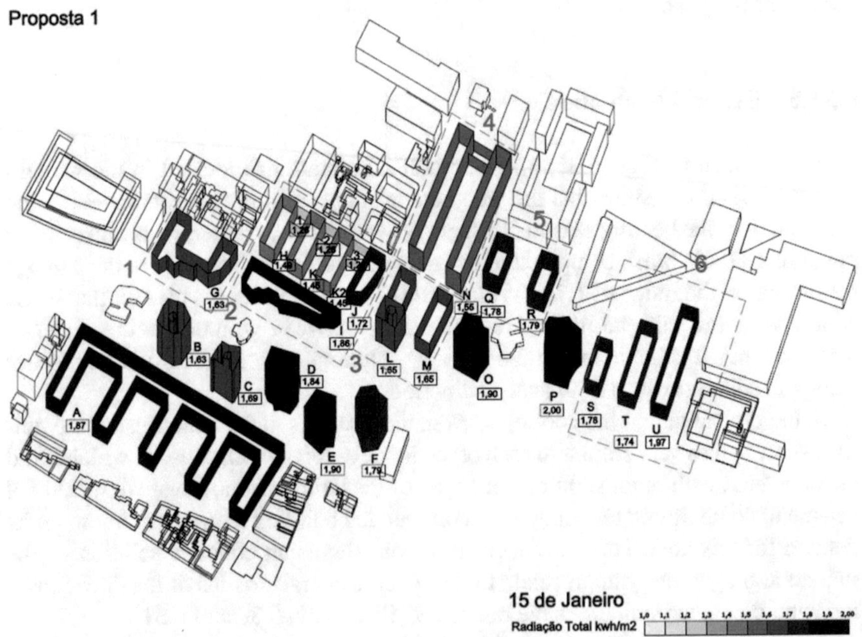

Fig. 6.27 Overall analysis of proposal 1 for the typical day of the heating season

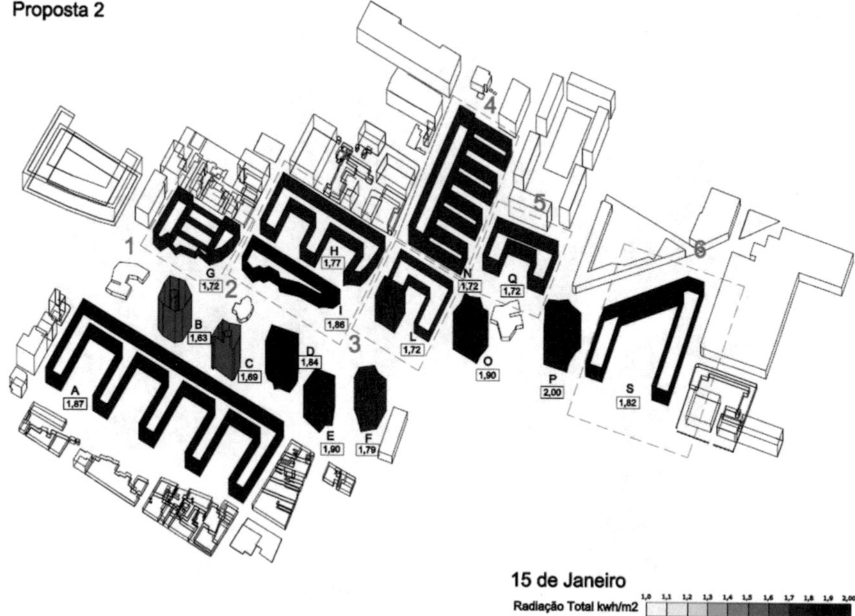

Fig. 6.28 Overall analysis of proposal 2 for the typical day of the heating season

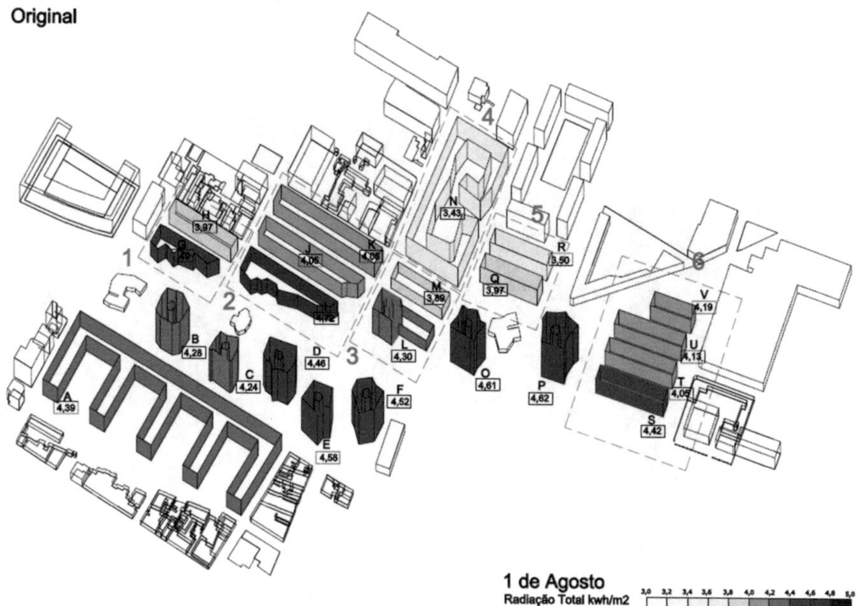

Fig. 6.29 Global analysis of the original scenario for the typical day of the cooling season

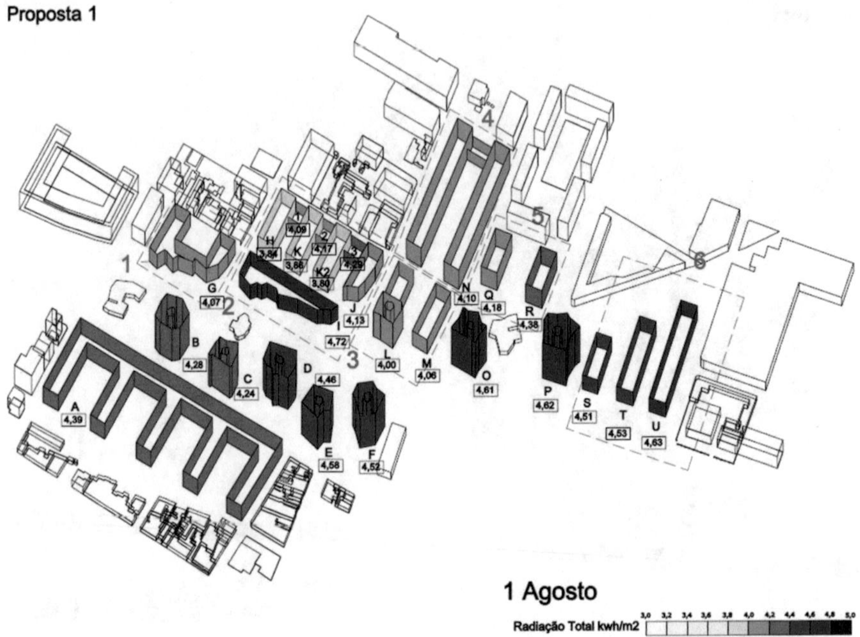

Fig. 6.30 Overall analysis of proposal 1 for the typical day of the cooling station

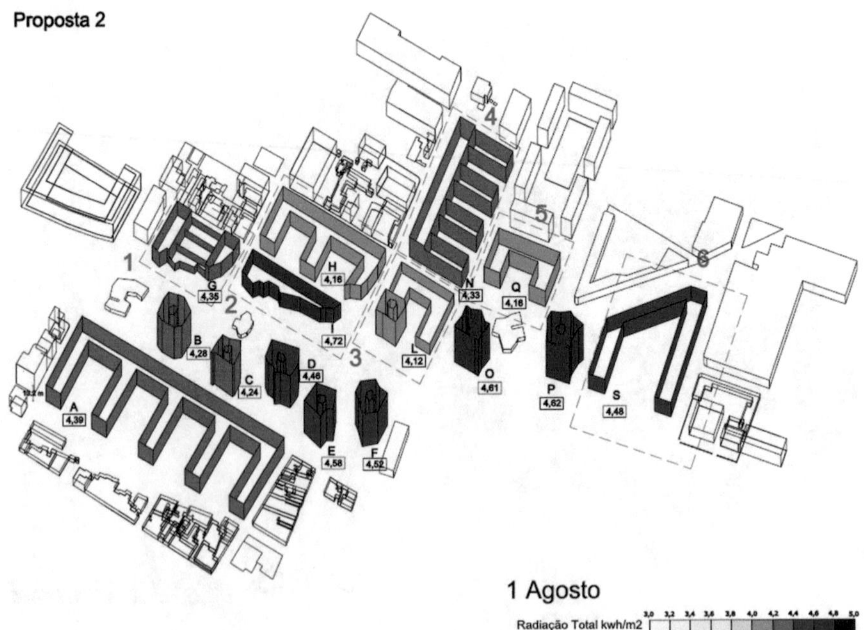

Fig. 6.31 Overall analysis of proposal 2 for the typical day of the cooling season

6.3.5.6 Final Considerations

Analysed in the different urban solutions with the same construction area and maximum heights, we understood that there is no exceptional and successful unique model of urban solution, but better individual solutions for each unit analysed in both urban solutions and model proposals.

Of the six units analysed, the best results were divided into equal numbers by proposal 1 and proposal 2, (3 in each one).

The most relevant conclusions are:

- The original model, in addition to presenting the highest form factor among all solutions, does not obtain satisfactory results in any of the analysed units, always obtaining the worst result among all.
- The solutions with the orientation of the facades with greater area of exposure to the west and east sides are always adverse, except when there are volumes that can guarantee the shading in the summer, although with clear damages in the winter.
- The solutions with greater frontage area exposed to the south are clearly the best and can result in interesting U-solutions when the proportion of facades to the south and north are superior to the facades facing east and west.
- Urban solutions deemed beneficial for obtaining solar radiation in winter were also the best solutions to ensure a lower incidence of solar radiation in the summer.

6.4 Conclusions

This chapter provided a general overview of the main issues concerning urban and building design in seaside areas in Portugal. It presented brief historical evolution, followed by aspects of urban regulations on environmental protection issues, the impact of tourism and best practice strategies for sustainable urban and building design. Case studies were presented, at urban and building scales.

It was shown that, both in terms of legal framework for urban design and it terms of knowledge of appropriate sustainable building solutions, the tools exist to provide effective answers to the main problems concerning coastal areas – namely, the growing human pressure and, on the longer term, sea level rise.

Portugal is a coastal country, and in line with the global trend of occupation of littoral areas, most of the population lives near the sea. The sea has moulded the history of the country for centuries, from the maritime discoveries to international trading, providing wealth and food, and still today has an enormous impact on the country, for example, through tourism.

Most of the problems related to the growing urbanization of the coastal areas have so far being resolved, with a certain degree of success. The entrance of Portugal in the EU boosted concerns on environmental issues, and today large portions of the coast are natural reserves, where new construction is strictly forbidden. The POOC regulation for coastal areas also forbids new construction less than 500 m from the seashore. Thousands of illegal settlements were eliminated, giving place to natural areas.

But these are minor challenges when compared to the problem of the sea level rise. The scale and impact of sea level rise, resultant from global warming, require a radical exchange in international policies, in worldwide living paradigms. Unfortunately, these changes are taking a long time to be implemented…. On a more positive note, extraordinary examples of resilience and adaptation can be found throughout the history of the country: from the brave resistance against Romans, Spanish and French to the incredible maritime voyages where new worlds, new cultures, and new ideas were discovered….

Chapter 7
Climate Adaptive Design on the Norwegian Coast

Luca Finocchiaro

7.1 Introduction

Coasts represent the meeting point between land and sea, giving access to resources from both the sides. Besides the exchange of goods, coasts make it possible to exchange information between different countries, playing a fundamental role for both the economical and the technological development of a territory. The Norwegian coast, extending itself for over 2500 km on the northern hemisphere, did not only play a fundamental role in the socioeconomic development of the country, but it is also responsible for the distribution of different climatic regions throughout the Norwegian territory (Fig. 7.1). Climatic regions throughout the country determined not only the distribution of flora and fauna but also the possibility for humans of inhabiting specific areas. Besides this, as Karl Otto Ellefsen observes, "it is not possible to recognize a proper tradition in coastal architecture in Norway." In constructing buildings along the coast, people generally adopted the same principles used on the inner land. Moreover, their use was limited to the fishing season. However, in a few areas of the country, site-specific solutions have been developed in order to adapt to the specific climatic conditions of the Norwegian seaside. In this chapter, solutions in historical and contemporary case studies are analyzed and then discussed in connection to recent climatic phenomena related to global warming.

L. Finocchiaro (✉)
Faculty of Architecture and Design, NTNU – Norwegian University of Science and Technology, Trondheim, Norway
e-mail: luca.finocchiaro@ntnu.no

A. Sayigh (ed.), *Seaside Building Design: Principles and Practice*,
Innovative Renewable Energy, https://doi.org/10.1007/978-3-319-67949-5_7

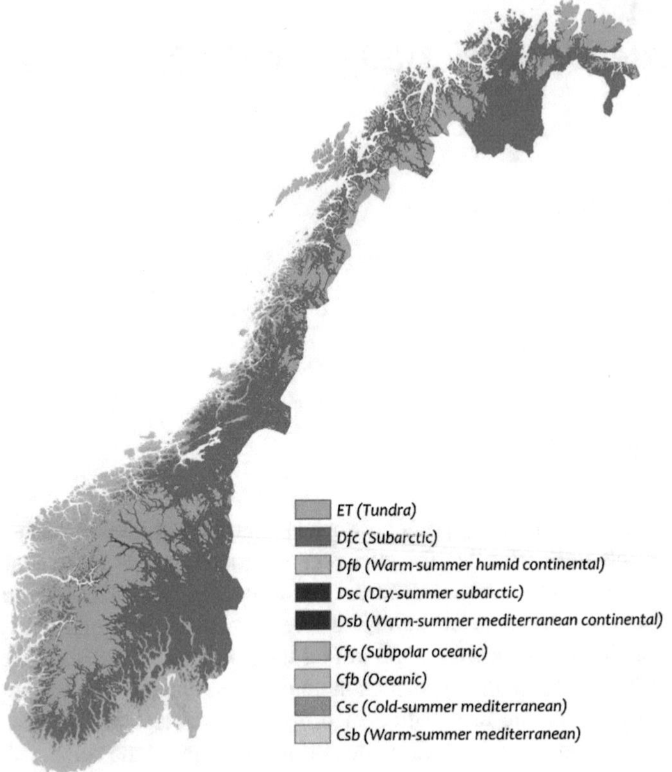

Fig. 7.1 Climatic chart, according to Köppen Geiger classification, (Source: Adam Peterson. Data extracted from WordClim.org)

7.2 The Norwegian Seaside

Throughout history, the Norwegian coast played a fundamental role for the socio-economic development of the country. Vikings were nomad warriors spending large part of their life on the sea, establishing villages in those parts of the coast where climatic conditions were still favorable for farming activities (Herje and Høyem, 1994). The sea did not only represent for them a fundamental way for the exchange of goods but also a way for exploring unknown territories beyond the European boundaries. Archaeological rests, found in the northern part of Norway, gave evidence that, throughout the middle age, climatic conditions were more favorable for farming than it is today. Vikings were able to cultivate oats and barley for their sustainment or flax and hemp for the production of textiles.

Historical documentations and archaeological rests gave evidence of a rich fishing trade between the north and the south of the country already in the eleventh century. The economy of Bergen flourished in those years as part of the Hanseatic alliance, working as an important hub for the exchange of alimentary goods between

Fig. 7.2 The "brygger" in Bergen (Photo: Dag Nilsen)

Norway and Europe. "Brygge," or ware houses,were raised in those years along the eastern side of the Vågen harbor in order to store stockfish from Lofoten and cereals coming from Europe (Fig. 7.2).

During the middle age, the Norwegian population sprawled throughout the territory, looking for those fertile parts of the land able to provide sufficient fruits for their sustainment. Because of the complex orography of the country, only a rather small part of the Norwegian territory was suitable for farming activities. Life was not easy, and, in a few cases, people had to settle down in quite harsh conditions, building houses and farms on improbable locations, uneasily accessible and in isolation from the city. Such difficult living conditions inspired, in the nineteenth century, a national romanticism, in literature and art, where daily activities related to life and farming are represented in a powerful and stunning nature. As Tora Karoline Arctander notice in her analysis about coastal architecture, conducted in 2009, besides the tight connection with the sea, activities on the seaside were generally not considered as being part of a regular family life. Fishing activities were in those years a less common practice than are today, often run by wealthy families owning entire villages in north Norway. Difficult environmental conditions due to the exposure to strong winds and rain did not encourage life along the coast. Buildings on the coast were therefore used during the fishing season only, when a large number of cods coming from the ocean would deposit their eggs along the coasts of Lofoten (Fig. 7.3).

With the industrialization of fishing activities, economic poles throughout the coast, such as Svolvær, started to flourish. This attracted a larger community of

Fig. 7.3 Interior of a traditional rorbu, a residential unit for fisherman, along the coast

people to settle down for the whole year along the Norwegian seaside. Industries fostered urbanization phenomena, making it possible for people to gradually delegating farming activities to the domain of larger cooperatives. Today circa 80% of the Norwegian population live at a distance of maximum 10 km from the seaside (Miljøstatus). With the transition of the fishing industry from a family-driven activity to the domain of large companies, fishing villages have transitioned into a new era where most of their building stock has been converted into housing or devoted to the tourism industry (Fig. 7.4).

7.3 Norwegian Coastal Climate

Tectonics of coasts represent the result of millions of years of evolution in which environmental forces such as water and wind shaped rocks and earth to their convenience. In Norway, waves from the sea and water from the valleys eroded the stones creating the most astonishing landscapes. Its irregular coastline measures over 2650 km and is surrounded by a multitude of islands. The Scandinavian Mountains, running from the north to the south of the country, quickly rise from the sea level to heights of up to 2000 m at a latitude proximal to the Arctic Circle. Their geology is the result of millions of years of uplift processes. Seawater enters in the inner land of the country through fjords up to 200 km long.

Fig. 7.4 Nusfjord, a fishing village along the coast of Lofoten, today converted into a touristic pole

The complex orography of the Norwegian territory is responsible for a highly variable set of environmental conditions (Fig. 7.5). High mountains and deep valleys, proximal to each other, are responsible for large differences in temperature, precipitations, and wind patterns even at a few kilometers of distance. For this reason, the Norwegian territory counts the largest number of natural habitats in Europe.

Because of its vertical extension on the northern hemisphere, spanning from 59° to 71°, seasonal differences are also markedly different in Norway. This is mostly due to the different contribution of the sun in terms of radiation. Differences are therefore larger with the increasing of the latitude. In north Norway, days are, for instance, 24 hours long during the summertime; in winters, on the other hand, a diffused light coming from south for a few hours creates astonishing colors in a joyful play with water and snow.

Besides the fact that Norway is extended mainly in the vertical direction on the northern hemisphere, the most remarkable differences in climatic conditions can be found when moving from the west to the east side of the country. Climatic differences are mostly due to the interaction between the mass of the ocean and the Scandinavian Mountains, running all along the Norwegian coast. Their presence is able to stop storm fronts from the ocean, on the west side, and continental climates from the east. Humid masses of air coming from the ocean are quickly lifted up by the mountains and cooled down, determining a larger amount of precipitations on

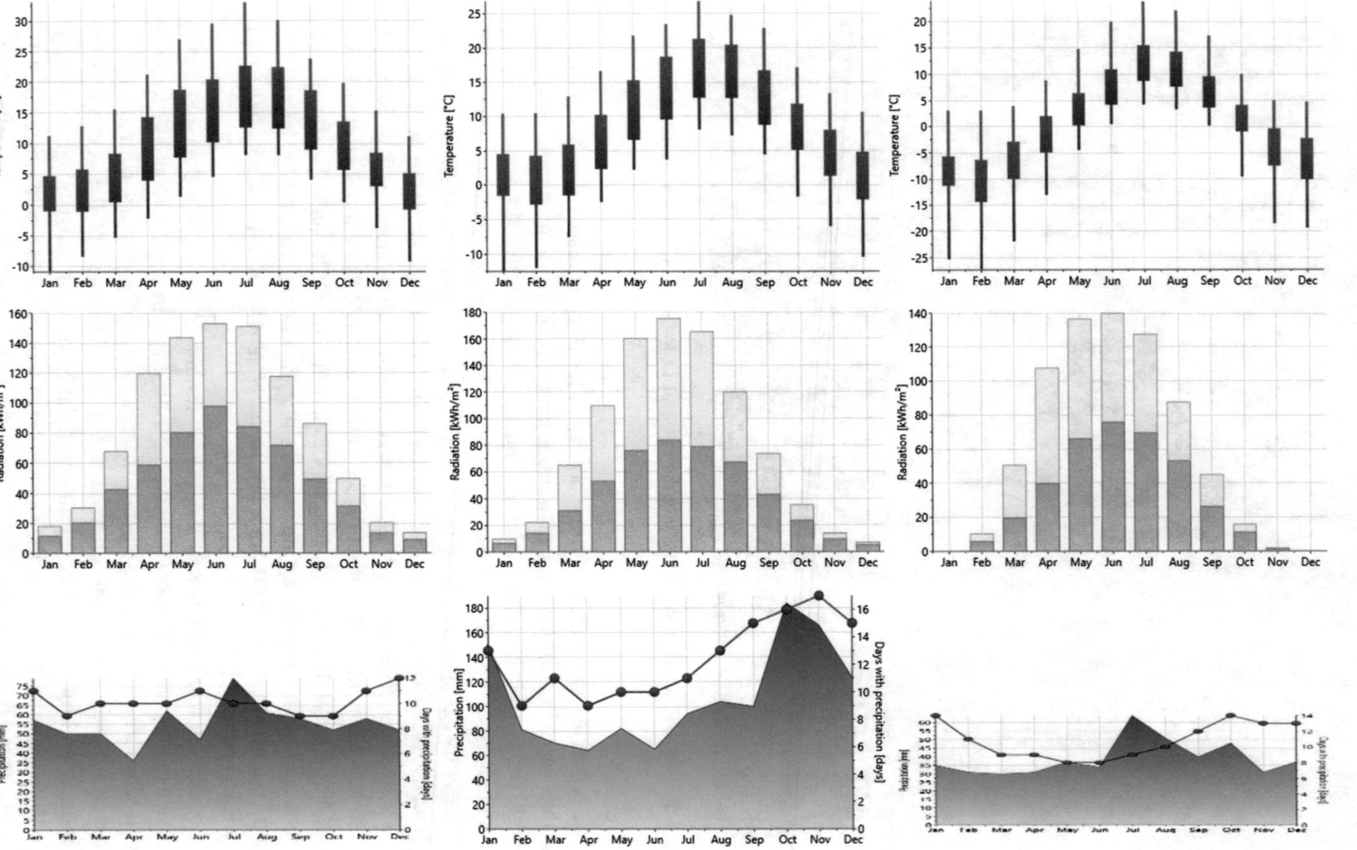

Fig. 7.5 Climatic data comparison for Bergen, Kristiansand, and Kirkenes (Data source: Meteonorm)

the coast than on the inner land. The city of Bergen collects on average 2250 mm annually, while coastal regions around Oslo are in comparison much drier with around 763 mm (YR, 2017). The Scandinavian Mountains do not either let continental mild winds coming from the east side reach the coastline. Because of this, summers throughout the Atlantic regions are generally cooler than on the inner land.

Climatic zones traced according to the Köppen-Geiger classification on the map of Norway show therefore three long stripes parallel to the western coast, belonging to the groups of temperate, subarctic climates and tundra – C, D, and E, respectively. The coast – group C – is generally characterized by a milder and temperate climate, when compared with inner lands and mountain areas. Coastal areas are dominated all year round by the polar front, leading to changeable, often overcast weather. Summers are cool due to cool ocean currents, but winters are mild and cloudy. Tundra – group ET – occupies the northeastern coast of the country or those parts of the mountains above the tree line, while inner lands are characterized by continental climates with warm summers.

Temperate oceanic climates – *Cfb* – cover most of the Norwegian coastline. These regions are exceptionally mild when compared with areas on similar latitudes elsewhere in the world. Mild temperatures are due to the Gulf Stream coming from the Atlantic coast and reaching the northern part of the country. This makes it possible to avoid ice-covered areas in the wintertime, for large part of the west coast, from Lista to the Lofoten islands. Subpolar oceanic climate (Cfc), characterized by larger snowfalls and milder winters than subarctic or continental climates, can be found close to the coasts of southern and western Norway or, in Northern Norway, even at altitudes proximal to the sea level. In the coastal areas of the southern and western part of Norway, annual average temperatures can be as high as 7.7 °C. Summers can be there warm and humid. Those areas are therefore generally classified as humid continental climates – Dfb.

7.4 Climate Adaptive Design on the Norwegian Coast

The oldest fishing settelments disovered on the Norwegian seaside, in the region of Hjartoy in Øygarden, date to the eight century and included 36 houses and 9 boathouses. During the fishing season, these houses could accommodate a community of 160 fishermen coming from farms sprawled throughout the country. With the refinement of the fishing industry, villages on the sea line started to include a whole series of different building typologies, reflecting the variety of activities required for running the whole industry. "Rorbuer" were built along the coast with the purpose of hosting fishermen during the fishing season. Larger industrial buildings started to be diffused with the advent of machineries from the mid-nineteenth century. On the other side, warehouses, or "brygge" in Norwegian, were built already from the eleventh century in several coastal cities with the purpose of storing and exchanging goods coming from all Norway and beyond.

Fishing activities presumed the construction of houses and storages in those parts of the coast where boats could more easily access. However, those areas were often exposed to extreme climatic conditions typical of the open landscapes of the sea line. According to what Karl Otto Ellefsen observes, it cannot be identified a coastal architecture tradition in Norway and characteristics of coastal buildings are often dictated by functional reasons (Ellefsen 1985). Most of the buildings on the seaside were constructed following the same principles as the ones on the inner land. Only the village owner's house, in the desire to express wellness, recurred to architecture styles in vogue throughout Europe or was equipped with state-of-the-art technologies. These were in most cases features merely related to the building's aesthetics and, only in a few cases, to the need for adapting to the specific climatic conditions of the seaside.

Besides this, a few elements can still be recognized in the architectural scale in the effort of adapting to seaside climatic conditions. Surfaces exposed to prevailing winds, identified as "værvegger" or weather walls, were generally built without any window in the attempt of minimizing infiltrations through the envelope. In a few cases, such as it happens in old farms in Averøy, those walls were built out of stone or bricks, with the purpose of minimizing their more demanding maintenance. Morphological characteristics of "jærhuset" (Fig. 7.6), detached houses in the southeast coast of Norway, were optimized in order to minimize the stress coming from the wind. For the same purpose but recurring to a different approach, "skuten" were particular versions of log houses where the wooden core was wrapped with a buffer space protecting the building core from prevailing winds. The buffer zone was used as a storage facility but also aimed to minimize air infiltration through the logs.

Whenever climate adaption is not solved at the building scale, compounds of buildings are placed in the landscape in order to better adapt to different climatic conditions. In the island of Grip on the west coast, a tight compound of buildings of even height – generally identified as *klyngebebyggelse (Schjelderup and Brekke)* – let the wind flow above the village creating a more comfortable space in between them. Buildings on the boundary are therefore more exposed to extreme climatic conditions and characterized by more aggressive facades, while houses open more toward the interior creating pleasant spaces where most of the village activities are run (Fig. 7.7).

Tundannelser (Schjelderup and Brekke) indicates also a small compound of buildings wrapping an outdoor area for different functional activities. This building typology inspired the open-air house designed by the architect Per Line on the white coast of Orrestranda in Jaeren (Fig. 7.8), in the southwest coast of Norway. The organic form gently sets itself in the open landscape. The low and aggressive walls toward the exterior create an aerodynamic form protecting an inner courtyard from winds and hard rain. Cafeteria and indoor spaces designed to accommodate exhibitions or simply for gathering people can be therefore extended to the inner courtyard for a large part of the year (Fig. 7.8).

Another interesting example of climate adaption on the Norwegian coast is the aquarium of Lofoten, lying along the coast of Vågankallen, one of the most astonishing geological formations and landscapes of the islands' archipelago. The archi-

Fig. 7.6 Morphological characteristics of a *jærhuset* were defined in order to minimize stress coming from the wind. A buffer space around the living area protected the core from infiltration issues (Photo by Hans Jacob Hansteen. Plan by Asgeir Bell)

tecture office Blå Strek i Tromsø recurs to traditional elements coming from the industrial heritage of Lofoten, such as circulation ramps or exposed structural frames as elements for giving the project the regionalist character of the islands. For this reason, the architects define the building as the result of critical regionalism. The aquarium includes two bodies: a public gallery covered by a pitched roof, including a cafeteria and exhibition space, and a closed body characterized by a curved façade toward the landscape, including most of the aquarium facilities. Circulating throughtout the building form, through ramps and terraces, people are able to connect themselves to coast and nature in several points of the structure. The curved façade of the aquarium does not only open to different views of the landscape but also wraps the inner space creating a form that gently plays with sun and wind. See Fig. 7.9.

Fig. 7.7 Grip. Houses scattered on the landscape avoid wind channelsto access the spaces in between the buildings. (Source: Rasmus Barstad)

7.5 Global Warming Phenomena on the Norwegian Seaside

Technological development of building materials made it possible nowadays to construct more resistant building components able to cope with adverse climatic conditions such as the one of the Norwegian seaside. This became the opportunity for localizing buildings more freely in the landscape, in a tighter connection with nature

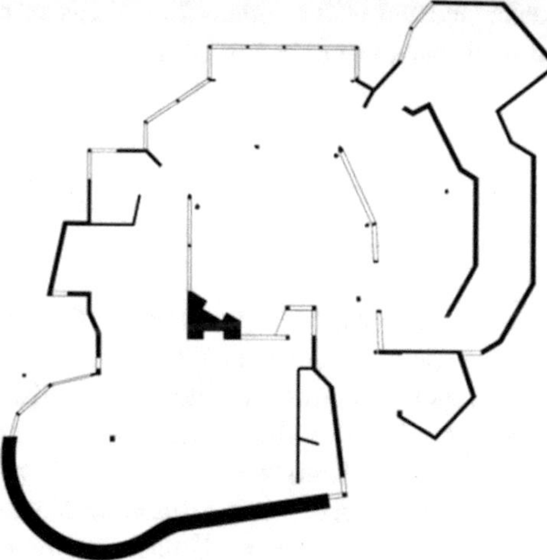

Fig. 7.8 The open house, designed by the architect Per Line

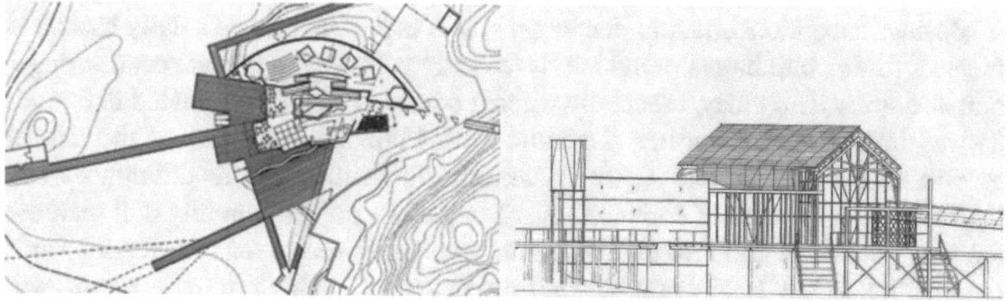

Fig. 7.9 The aquarium of Lofoten (Source: Blå Strek arkitekter). A curved façade minimizes the wind stress over the building volume while maximizing view toward the sea

and view. On the other side, however, phenomena related to climate change are rising a high degree of uncertainty related to the climatic stress to which the built environment will be exposed in the near future.

Data collected in the last 30 years show that, because of global warming phenomena, annual average temperatures throughout the country have risen between 1°

and 2.5° when compared with data collected between 1961 and 1990 (Klima 2020). For a large part of the coast, from Kristiansand to Svolvær, annual average temperatures are today, in fact, above the freezing point. Summers last longer than a few years ago, while winters are getting shorter and shorter. Snow cover has also tended to decrease over the mountains, while precipitations have generally increased in the mainland. As a consequence, the tree line has moved to a higher quote, letting a larger part of the territory to be covered with woods. For this reason, substantial areas of Norway can nowadays be classified in a different climate zone than when compared to the years between 1961 and 1990. "Oslo and Trondheim could for instance be classified as maritime temperate (Cfb), Tromsø as cool maritime (Cfc), and Lillehammer, earlier located at the intersection between subarctic (Dfc) and humid continental (Dfb) climate, would be firmly humid continental" (Klima 2020). See Fig. 7.10.

7.6 Conclusions

The complex orography of the country in proximity of the coast, in combination with the exposure to strong winds coming from the ocean, is historically responsible for extreme natural events such as storms, avalanches, or landslides. Increased temperatures and precipitations related to global warming, especially in autumn and winter, already increased erosion phenomena of mountains' rocks and the risk of landslides. According to meteorologists, such extraordinary natural events are destined to be more frequent in the near future. Global warming will result in even more extreme climatic conditions and put inedited challenges related to climate adaption on the different building typologies (Klima 2020), causing severe damages to the built environment.

Besides exceptional natural phenomena, it is expected that, on a daily basis, "in the near future, buildings throughout the country will need to tolerate even stronger climatic stress than today, especially when it comes to challenges related to humidity" (Klima 2020). A warmer and more humid climate will increase the risk of growth of mushrooms and algae. In combination with pressure differences and hygroscopic properties of commonly used materials, higher humidity will increase the risk for leakages and rotting of internal wooden elements such as wooden studs or structural frame. Heavy rain, driven from fast winds coming from the ocean, will result in increased stress on the vertical surfaces and joints of the exposed parts of the envelope (Lisø and Kvande 2007).

Inedited boundary conditions will soon require a new effort for the development of technical solutions for ensuring an optimal performance of the building skin. Climatic stress could affect building functionality and the whole building life cycle. For this reason, a set of climatic charts have already been developed within the Klima2020 project in order to provide a platform for understanding the grade of risk connected to future climate stress. Charts were developed in order to give evidence of an increased risk of rotting of wooden elements and give a first estimate of the

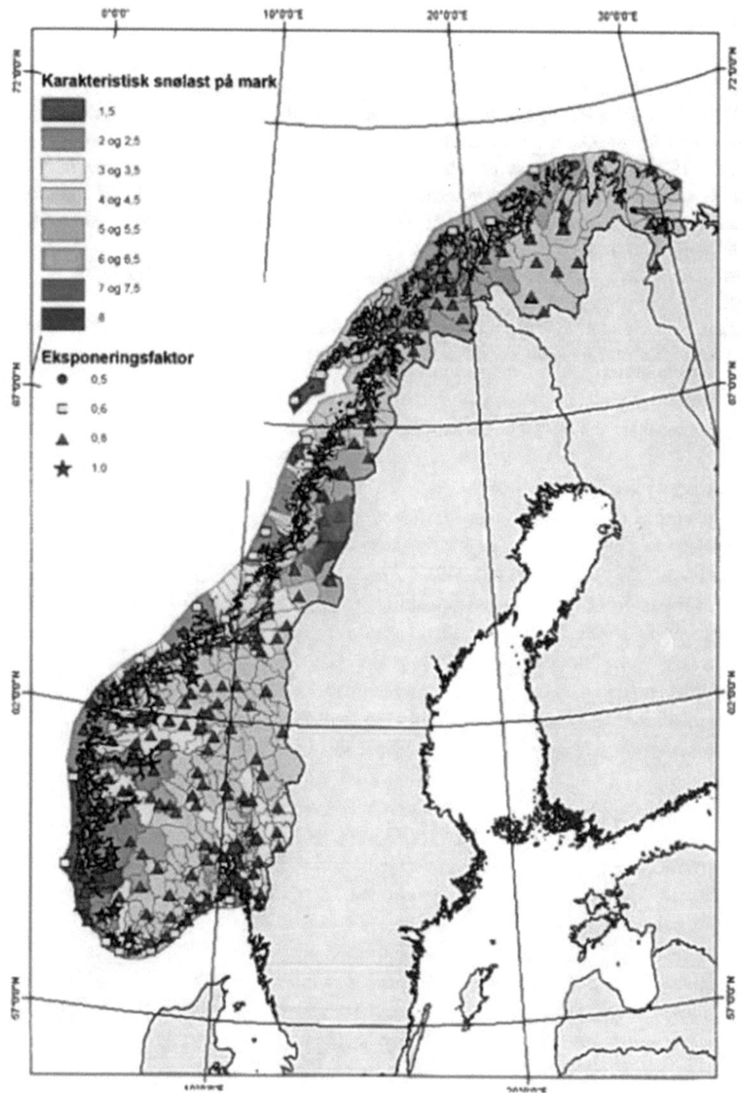

Fig. 7.10 A risk map highlights new boundary environmental conditions along the Norwegian coast (Klima 2020)

interaction between building materials and inedited boundary conditions. Risk analyses become in this regard a tool for the architectural design of climate-adaptive buildings. Besides this, architects will need to develop the ability of developing their own solutions, using site-specific climatic data and indexes in order to define innovative solutions to apply in each specific project. Intensity of precipitations, wind direction, and speed must all be taken as design parameters in the effort of reharmonizing the natural and the built environment.

References

Ellefsen, K. O. (1985). Fyskeværet. *Den norske Landsbyen*. Byggekunst 7, p. 383.

Herje, J. R., & Høyem, H. (1994). *Vind og Vær: Håndbok i klimatilpassing av bebyggelse i vindutsatte strøk i Norge*. Oslo: Husbanken.

Høyem, H. (1997). *Coastal, location and site*, 1.III. Cambridge: Cambridge University Press.

Lisø, K. R., & Kvande, T. (2007). *Klimatilpasning av bygninger*. Oslo: SINTEF Byggforsk.

Schjelderup, H., & Brekke, N. G. (1997). *Hus på vestkysten gjennom 4000 år*. Bergen: Fortidsminneforeningen.

Tora Karoline Arctander, Kystarkitektur. Master thesis in Architecture, AHO, 2009.

yr.no/.sted/Norge/Hordaland/Bergen/Bergen/statistikk.html

Norwegian Environment Agency (2016), Coastal Waters. Retrieved from: http://www.environment.no/topics/marine-and-coastal-waters/coastal-waters/

Chapter 8
Green Design for a Smart Island: Green Infrastructure and Architectural Solutions for Ecotourism in Mediterranean Areas

Antonella Trombadore

8.1 Green Design Approach in Mediterranean Areas: Five Ideas to Share

Why do we need to become green? What is the design approach for green building in Mediterranean cultural and climatic condition? Is it green in color? Is it a building covered with plants? Should it be in a brownfield site? Is it a high-tech building?

Green building is the practice of increasing the efficiency of buildings and their use of energy, water, and materials and reducing building impacts on human health and the environment, through better site location, design, construction, operation, maintenance, and removal, taking into account every aspect of the complete building life cycle. What is the "green challenge" in Mediterranean areas?

Five ideas to share (Trombadore 2015):

Architecture and Climate Condition The Mediterranean climatic condition requires building appropriate solutions: the problem of energy consumption for summer comfort cannot be solved by following the logical construction of Northern Europe. To reduce energy cost, it is necessary to define innovative strategies in the building, which is strongly related to climatic and cultural characteristics.

Inclusiveness and Change The fast and dynamic evolution of social and demographic structure of the population of the Mediterranean area suggests a change from the past that determines the need for new models of urban spaces and residential use, with typological and technological innovations to support new social and intercultural issues.

A. Trombadore (✉)
ABITA Interuniversity Research Centre - DIDA Architectrual Department, University of Florence, 50125, Via San Niccolò 93, Florence, Italy
e-mail: antonella.trombadore@unifi.it

A. Sayigh (ed.), *Seaside Building Design: Principles and Practice*, Innovative Renewable Energy, https://doi.org/10.1007/978-3-319-67949-5_8

Identity and Competitiveness The Mediterranean ecosystems and environmental assets are based on the knowledge of the art of living, the management of the landscape, and the ability to stimulate cooperation on issues of sustainable development. The regional environmental identity is based on architectural elements and cultural climate in which the distinctive scientific subjects, companies, and governments should aim to boost competitiveness at international level.

Urban Transformation and Environmental Quality Forced transition to a sustainable city of strategic importance to revitalize the area. To manage the population growth in the Mediterranean, resulting increase in energy consumption, and increased demand for comfort, it is necessary for the construction market that provides appropriate solutions to new social needs and more environmentally friendly technologies.

Innovation and Tradition The Mediterranean architectural traditions of the past, full of potential and interesting cultural influences, represent an important heritage of civilization, and it is from these that we must be inspired to develop new building components with high-energy performance. There is the need to recover and innovate traditional design principles and increase the level of environmental and architectural quality of the interventions of building retrofitting and new construction (Figs. 8.1 and 8.2).

8.2 Ecotourism as Territorial Challenge for Small Islands

In the Mediterranean regions, tourism has always played a strategic role with significant territorial impact in economic and ecological terms, but over the years there has been a strong gap between coastal seaside tourism and cultural and scientific tourism. This phenomenon, similar in all the islands albeit different in dynamics, resulted in the 1970s and 1980s exodus and demographic impoverishment of the smaller islands, with consequences on the management and maintenance of housing stock, on the economic activities of the present historic towns and the abandonment of the settlements and the countryside: a depletion of resources, reducing the competences and services that dramatically lowers the tourist attractiveness of the territories, even if placed in prestigious contexts. In the Mediterranean area, you can identify at least three different types of smaller islands, with a long and troubled historical sedimentation, and today almost all of these islands are considered protected natural areas, whose management and protection are entrusted to national or regional parks. Small islands are still full of stories of treasures, memories, and symbols that have undergone a process of transformation in recent years, often uncontrolled and characterized by fragmented management, which altered the socioeconomic characteristics, the land use, and the use and regeneration potential.

Figs. 8.1 and 8.2 Typical urban aspects of the Mediterranean island

The Asinara island, located in the north-west of Sardinia, includes the National Park of Asinara and has a surface of 51.9 square kilometers, formed by four minor mountain systems that are surrounded and linked by a narrow and flat coastal belt.

The marine and underwater island environment is a protected natural area, and the island, in its territory, represents a uniqueness in the Mediterranean area from a geological point of view, as well as its flora and fauna, both on land and sea, given the state of its territorial naturalness and outstanding presence. The island of Asinara in different centuries undergone alternation of cultural influences, creating a succession of different social dynamics of communities and settlements: from Cala Oliva as the village of Ligurian fishermen to various hospitals in the first and second period, as well as Cala Reale and the agricultural structures of Campu Perdu and Santa Maria and the prison sectors of Trabuccatu, Fornelli, and Tumbarino. Despite the diversity of the social structure, the islanders have allowed us to maintain a dynamic balance between natural and man, being able to manage the entropy of the system in all its complexity, ensuring in parallel the development of agricultural activities in the economic and functional the island and the island life. The establishment of the Asinara National Park has fortunately slowed the phenomenon and led to a reversal of the trend, aiming at stimulating a growing demand for high environmental quality services, more aware of the rural and natural resource ecosystems, with a scientific approach and an international feel. Asinara is a privileged place to promote an innovative and sustainable tourism model for the territorial and regional context, where natural potential and resources related to the authenticity of the places they are dominant, but not adequately exploited in the logic of a responsible tourism Parco Nazionale dell'Asinara.

8.3 Clusters and Pilot Projects: Self-Sufficient Energy Islands Within the Island

The strategic plan designed to initiate a rebirth and Asinara island regeneration process, through the development of functions compatible with the fragile ecosystem, is based on green retrofitting actions of existing building, according to an eco-compatible approach and energy efficiency strategies integrated with the environment. However, to predict and to best design green buildings and to manage these new activities, it has developed an additional level of planning: the island's division into the clusters. The term cluster is redefined in this case and applied to the particular natural and cultural context in which we find ourselves. It is to divide the island into several parts, which are considered economic and energy self-sufficient. This brings us to the concept of the islands in the island: a complex of buildings, structures, agricultural, and natural areas full of all types of services and activities related to its function (Manuale Alcotra 2013).This self-sufficiency is related to one (or more) particular activity, well defined, with a series of facilities dedicated to it. These structures are usually obtained from the preexisting buildings. Each cluster has a main theme, and in some cases we have an additional division, as some activities are located in various parts of the island (Fig. 8.3).

8.4 Green Strategies and Green Infrastructure for a Nearly Zero Energy Island

The master plan is integrated with strategies for sustainability: it is important to reaffirm the peculiarity of the environment and the natural balance of the plan. The inclusion of new features, therefore, must not only be compatible from an

Fig. 8.3 General view of Asinara island in Sardinia, Italy

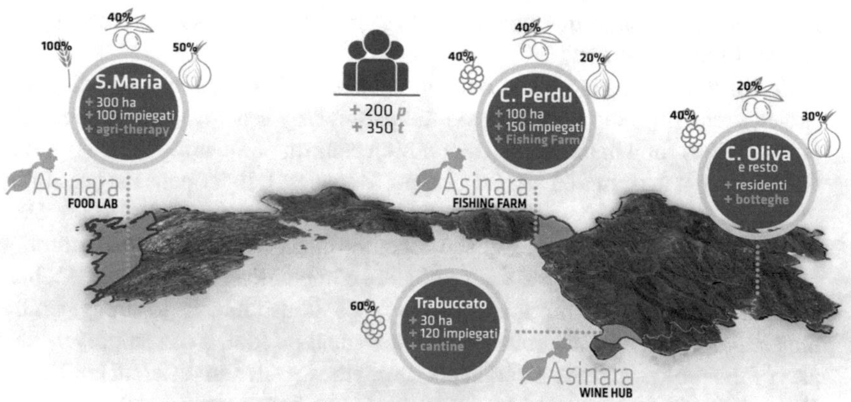

Fig. 8.4 The location and activities of the cluster in the master plan

ecological point of view, economic and social, but must be accompanied by a relevant resource management in an intelligent and forward-looking way. As for the identification of intervention strategies, the method adopted was to create a schedule of references, later adapted and redesigned to be part of the strategic plan. These solutions are based on a number of different possible interventions, ranging from energy efficiency field to the intelligent management of the waste and the organization of mobility (Fig. 8.4).

- *Water.* Water is one of the most critical resources within the insular, because the whole system is based on the management of a network of reservoirs that provides water for all uses on the island. The strategies implemented also, therefore, will focus on this existing system. It is expected, in the first place, a total recovery and restoration of tanks, with particular attention to those relevant to urban areas, as provided by the park plan.

- *Energy.* The second resource planning is the subject of sustainable energy. Compared to the situation with the management of water, electricity is in much better condition. Electricity is supplied to the island by an underwater cable along approximately 4 km; it continues along the main road, extending along the entire island. It is in a satisfactory condition, and the plan provides a parts repair and recovery infrastructure in order to obtain suitable to absorb the new features and their needs. The strategic plan foresees the inclusion of some structures and solutions for additional energy savings and renewal of resources. Shall refer, in particular the use of photovoltaic materials, but in an alternative way and updated to take over in the minimum visual impact fees that are to be respected.
- *Waste management.* The collection, waste management, and disposal are very interesting topics within the strategic plan. Although the plan of the park identifies the current system as functional and satisfying, the inclusion of new features and new structures creates an increase in tourist flows that could pose serious problems for the collection and management of waste. The plan provides an expansion of facilities suitable for the collection of waste and testing of an *anaerobic digestion* system that could turn organic waste into heat, electricity, and compost for agriculture.
- *Mobility.* It also plans to incorporate in the plan a vision and a completely new strategy regarding the island. Given the particularly sensitive environment, the plan provides for a maximum CO_2 emission control, eliminating the transport in the car, except for cases emergency, support services. It thus opens the possibility to also create a small network of public transport, of small size and 100% electric. Furthermore, it is expected that most of the displacement takes place through electrical machines, equipped with paths and complemented by different shelters and small facilities for information of all types. It also intends to strengthen the movement by bicycle and enhance the movement on foot, a concept that is also part of the project and the natural/sports activities related to different contexts.
- *Green infrastructures.* Green Infrastructures have been instrumental in architecture's quest toward greening its impact (ARUP 2014). These cross-disciplinary solutions have been used primarily on an urban scale, giving crowded and overbuilt cities a chance of creating green spaces, both on ground and inside, around, and on top of buildings. Using natural green elements as well as water, blue-green infrastructures have contributed to creation of spaces known as "city lungs." These solutions not only confront and diminish the negative impact of mass building but help in preventing and even inverting the pollution process that has been growing with recent urban developments. Some of the examples of green infrastructure applications are *green corridors*, which have multiple benefits, especially in reducing the *urban heat island effect* that has been devastating cities' climate systems recently. Islands can be seen as a perfect example of systems that are extremely fragile and sensible to climate change and that rely on intelligent and green planning in order to be able to function and contribute both socially and economically.

8.4.1 CanoPV as Green Infrastructures

The idea for *CanoPV+* arises from the need to create a special module to control solar radiation and improve the comfort in the external areas and public green spaces. The idea of the flower drives the basic concept design of the modules, as a petal that is combined by the simplest geometric shapes – triangle, square, and hexagon will produce the three typologies of CanoPV+. Three very organic forms were obtained, to replace the elements of local flora, integrating this module into the surrounding landscape. The three different sizes of the modules allow to obtain a shadow surface changing in the different daily or season times. The solution is very effective to collect rainwater; besides the integration of PV amorphous (between the double surface of the elements) will produce sufficient electrical renewable energy to cover the entire laboratory requirements and external lighting (Figs. 8.5 and 8.6).

8.4.2 Pilot Projects

The analysis of the island's environmental characteristics, the evaluation of its building's potential to be transformed into a green building, the choice of eco-friendly technological solutions, and the definition of pilot concepts design have been implemented during the international design workshop carried out in September 2016 at the island of Asinara. Thirty Italian students, in collaboration with Erasmus students coming from different European countries, attending the environmental design laboratory at faculty of Architecture in Florence, all together implemented the concept design for pilot projects. Many ideas have been subsequently developed as a thesis, fostering the analysis and performance evaluation of architectural and technological solutions. In particular the project *Santa Maria Green In Farm Structure* by Dusal Rolovic, *Euphorbia Lab as Mediterranean Phitorerapic Center* by Valentina Tavanti, *Campu Perdu Agri Farm and Biodynamic Agriculture Laboratories* by David Tinti, *Fornelli Cultural Hub* by Canio Telesca, and *Bio-Winery of Trabuccatu* by Giacomo Parrini.

8.5 Pilot Project N.1 Asinara Green in Farm Structure

The pilot project focused on eco-compatible solutions for the revitalization of the former prison sector Santa Maria as an International laboratory for ecotourism, compatible both with its touristic potential and especially with the natural fragility of the island, making extreme attention toward its cohesion with nature and biodiversity. This complex has been used in the past as a farm for a particular group of inmates and is based near a field of nearly 200 hectares that can be turned into agricultural factory. The concept behind the project was about creating a structure that

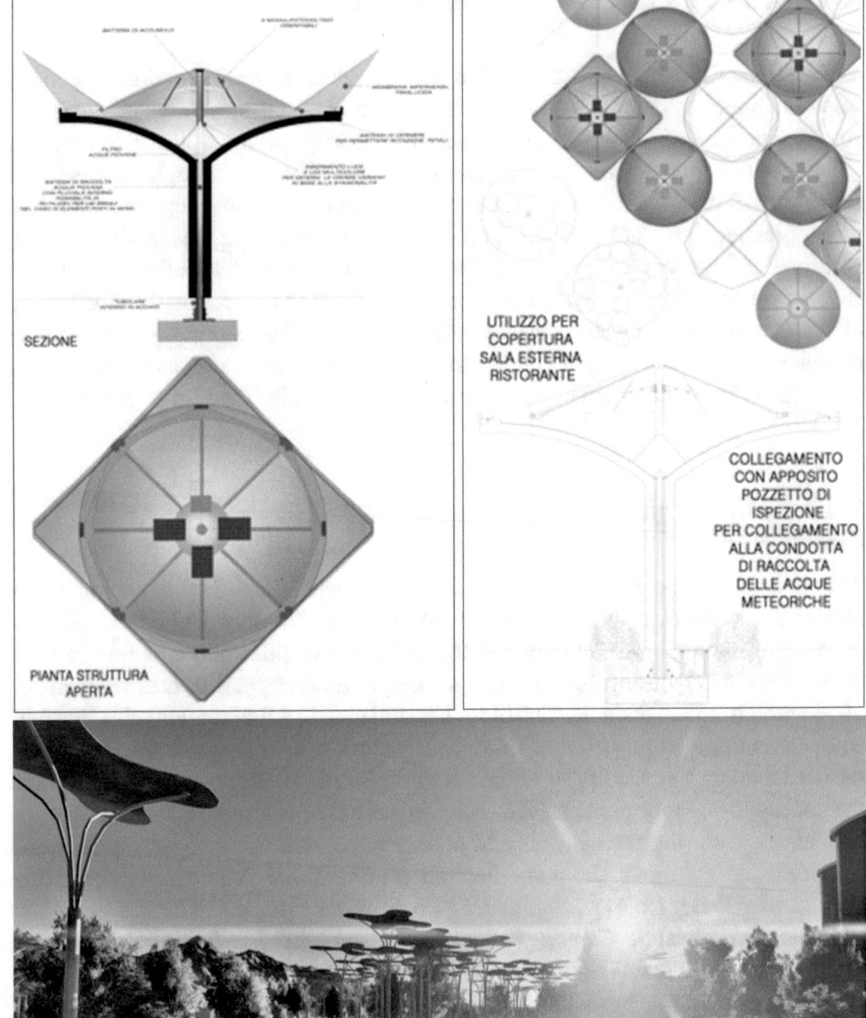

Figs. 8.5 and 8.6 View of the PV green infrastructures for renewable energy production

Fig. 8.7 General view of the pilot project

is both productive and commercial and that offers unique and naturalistic way of approaching and learning processes of agriculture but that offers beautiful and relaxing luxuries such as slow food restaurant, exclusive green suites, kitchen laboratory, and agri-therapy for children. It has been conceived as multifunctional set of buildings with green public spaces in between, perfectly connected and able to offer a full set of activities and leisure for tourists as well as researchers.

The building is a prison structure that has been revitalized and refreshed to house a slow food restaurant and various suites for visitors (Fig. 8.7).

The project leaves internal disposition practically intact but intervenes on buildings facades, making it much more open and illuminated, creating almost semi-open spaces that collaborate perfectly with the green surrounding them. These facades have been elaborated focusing on natural ventilation solutions and daylighting technologies, in order to reach a full-scale comfort inside the suites and restaurant. These technologies include:

- *EvaCool*, a separated façade system built with tube structure that envelopes plants and hygroscopic materials that help cool air via evaporation and channel it inside the building with specially designed cross-ventilation systems
- A special in-suite greenhouse that uses a skylight for water harvesting and daylighting, increasing the climatic comfort in the suite, indirect and diffused illumination
- Inside and outside *gardens* that house a terrace-based water harvesting system
- *CanoPV+*, a modular, tree-like canopy structure, with integrated photovoltaic cells, that serves both as energy-efficient element as well as for creating shaded spaces in outside gardens.

8.5.1 Green Room

It can be considered a nucleus of various technological solutions, with the scope of bringing perfect comfort and ideal thermal state inside the suites of the A Block. As suggested by its name, Green Room is characterized by a special in-suite greenhouse that uses a skylight for water harvesting and for better illumination, increasing the climatic comfort in the suite as well as indirect and diffused illumination. Standing between the in-suite bathroom and the bedroom, it has a wide array of functions. The greenhouse serves as a water harvesting system, with rainwater being collected from the roof, through the skylight, and to the vegetation. The humidity of the plants helps cool down the ambient and provides a unique experience with the bathtub in between the high vegetation.

The skylight at the top of the structure helps bring more natural light to the interior. With its glass structure and precisely inclined roof section, the Green Room helps diffuse natural light into the bedroom, giving more light in the morning period but maintaining low exposure during warm hours. The exterior of these suites has been enriched with a shading structure made primarily of wood and steel. It has a mobile front element that helps direct light throughout the day, contributing to the general comfort of the interior (Figs. 8.8 and 8.9).

8.5.2 EvaCool

A separated facade system built with tube structure that envelopes plants and hygroscopic materials. These materials help cool the air via evaporation and channel it toward the interior of the building with specially designed cross-ventilation systems.

The idea behind the creation of EvaCool is based on the need to create a new solution for the main facade of A Block, along with the intent of giving thermal comfort to the interiors. It is characterized by a tubular 3D structure with vegetation between the tubes. The tube system is connected to the green roof that collects rainwater and then channels it to the tubes. From there, part of the water is sprayed on the vegetation, and the rest continues to the main harvesting system. This way, hygroscopic materials in between the vegetation and tubular structure make contact with water, enabling them to instigate evaporation. The warm air gets colder and subsequently enters the interior via specifically designed openings above the doors. This system, combined with the cross-ventilation provided by aligned openings, assures thermal comfort inside the right wing of the building.

8.5.3 Comfort Room

The B Block houses another set of suites that have been designed with an effort to provide a complete thermal comfort. The Comfort Room is characterized by an external shading structure with mobile elements that provide shading as well as rain

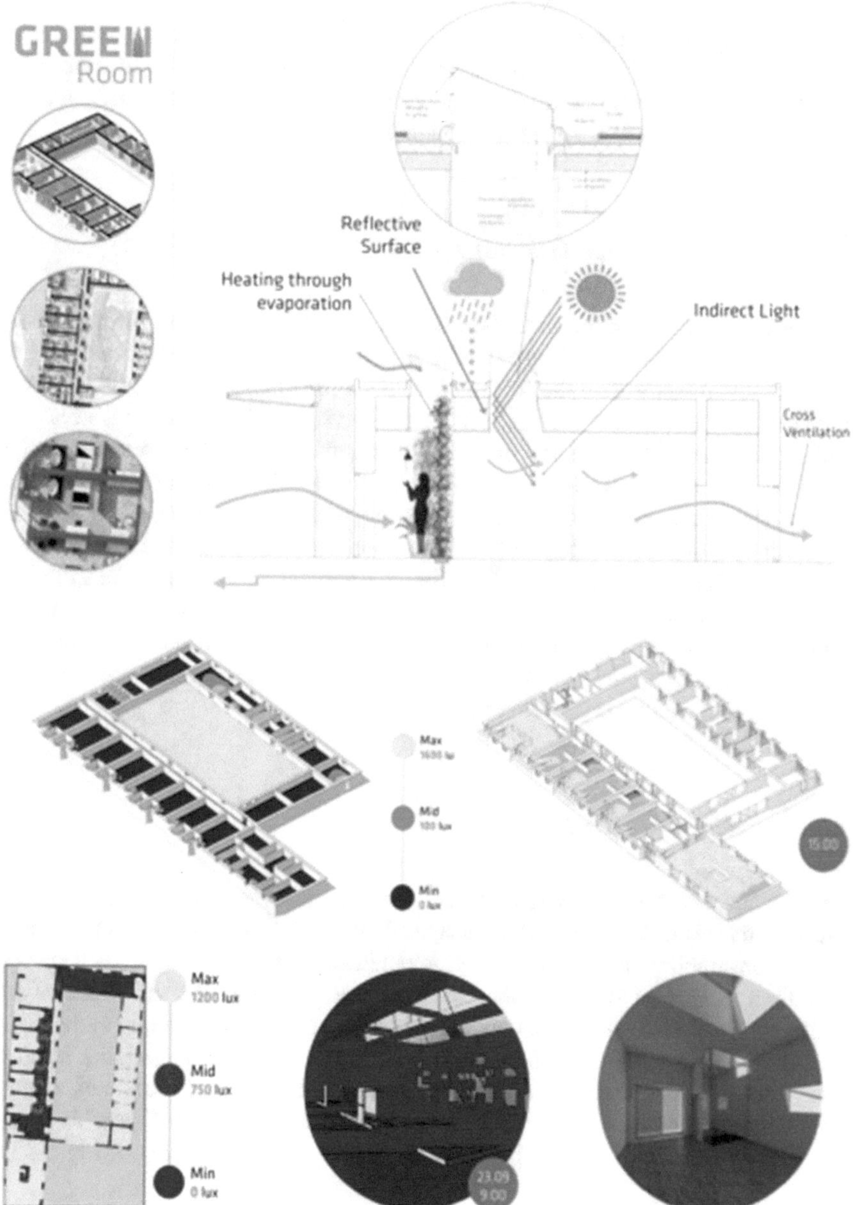

Figs. 8.8 and 8.9 Strategies of Green Room and daylight evaluation

protection and ventilation. Steel elements of this external structure is a part of general water harvesting structure that helps provide water for irrigation of nearby vegetable gardens. The original openings on the back side have been modified in order to provide more daylight and also to improve the cross-ventilation system, guaranteeing major comfort inside the suites (Figs. 8.10, 8.11, and 8.12).

8.5.4 General Water Harvesting System

All the abovementioned technologies are a part of a wider system for collection, depuration, and subsequent reuse of rainwater AD-Nett. This water harvesting system interacts directly with the two main gardens (one inside and one on the outer side) that house a terrace-based water filtration system. The rainwater is collected both from the roof structure and from the gardens themselves. The green roof helps clean the rainwater with its filter layers. The water continues and is conveyed through the EvaCool and Green Room Systems, where it has an important function for thermal comfort regulation. The excess water then continues and is attached to the underground system and conveyed through terraces inside small tanks that are integrated in the gardens. Water is then pumped from the tanks back to the building using energy provided by the CanoPV+ structure positioned in and around the gardens, contributing to the optimal eco-sustainable result, taking into account both water harvesting and energy efficiency.

8.5.5 Hybrid Heating System

Heating and warm water system is also designed to contribute to energy-saving goals of the project. It consists of a hybrid solution that minimizes the electric energy consumption and uses energy-efficient heat pump but also includes a special solar thermal system to help contain energy usage. The solar-thermal modules are positioned on the left wing of the green roof and interact with the ceiling system that houses all the installations. This system should contribute to minimizing electric energy consumption and can go as far as making it completely unnecessary during main summer months. Combined with the energy provided from CanoPV+ modules, this system further develops the eco-sustainable nature of the project (Fig. 8.13).

8.6 Pilot Project N.2 *Euphorbia* Lab as Mediterranean Phitorerapic Center

The Asinara Mediterranean Phitoterapic Center is thought to be an international hub where scientific research activities, production and processing laboratories, training, dissemination, and marketing activities will be integrated. The project aims at

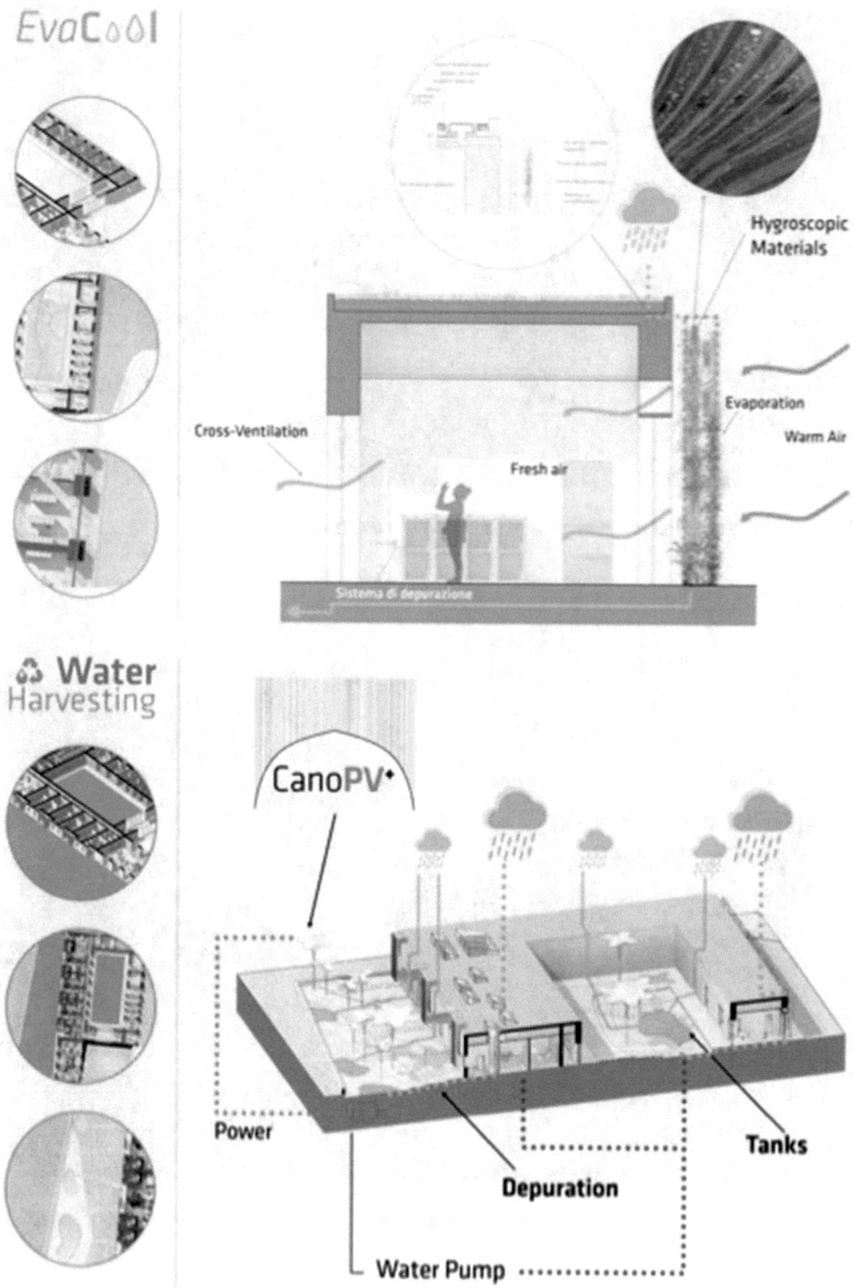

Figs. 8.10 and 8.11 Strategies of evaporative cooling and water harvesting

Fig. 8.12 The location and activities of the cluster in the master plan

Fig. 8.13 General view of the project: the environmental integration of buildings and of the green infrastructures

the realization of polyvalent laboratories where the active principles of herbs are extracted, with the opportunity to welcome students for live demonstrations. The interaction among research, didactic activities, dissemination, and wellness experience becomes the main goal of the project, promoting the participation of tourists and students in production/processing activities, offering them the opportunity to

Fig. 8.14 General view of the existing building selected to become the new Mediterranean Phitoterapic Center

have direct field experience, and creating accommodation spaces for students and researchers, as well as the realization of large greenhouses where herbs and plants will be stored. The research center is completed with an auditorium for 80 seats and a panoramic restaurant with km0 products. The aim is to create a multifunctional model, where the spaces dedicated to study, work, and relax can be interconnected, offering a real wellness experience.

The project for the recovery and enhancement of the building complex aims at experimenting ecologically friendly technological solutions, maintaining the balance between architectural volumes and surrounding landscape and enhancing the relationship between architecture and green: the restaurant will be surrounded by green walls and the upper terrace by the aromatic gardens, and the residences will be divided into cloisters and private themed courtyards. The different functions of the center are synthetically distinguished as follows: the workshop, the research center with classrooms and polyvalent laboratories, accommodation, residences for researchers and students, greenhouses for the cultivation of herbs and officinal herbs, tasting &sale, and wellness center (Fig. 8.14).

Following the approach of green buildings, the main aim of retrofitting action is to reduce overall impact on environment and human health by:

1. Reducing trash, pollution, and degradation of environment
2. Efficiently using energy, water, and other resources
3. Protecting occupant health and improving productivity

8.6.1 The Solar Radiation

The optimum orientation of the building complex, closely related to the conditions of healing therapies, has made it possible to optimize the most of the sun resource. The integration of the new volumes within thc existing wall perimeters was designed to enhance the "bioclimatic" features already in the system: the windows alignment allows the natural ventilation and the passive cooling, improving the indoor comfort during more sunny days. The great thickness of the wall allows to obtain an optimal thermal mass; the offset and the distance between the volumes allow each unit to capture maximum solar radiation. It is important to emphasize that the possibility of looking at large external spaces, especially if green, increases the level of inner comfort. The design of gardens and courtyards integrated to the residences enables the guest to enjoy nature and to experiment the maximum comfort in the ventilated microclimate of the gardens. What cannot be denied is the state of well-being linked to emotional feeling: like a beautiful panorama, the perception of the outside and, above all, the view of the sea can relax your eyes and stretch your mind. This is the purpose of a study looking into the layout of the internal volumes: each residential unit is offset from the backroom to allow to look outside. Playing on variable heights and alternating spaces, it was possible to create highly suggestive panoramic views, within each building, one on all the green promenade (Fig. 8.15).

8.6.2 The Water

It is well understood that the water resource has been and is still an important element in the design of the Asinara island area: once it is guaranteed the correct and sufficient water supply, we have to study how to obtain the most quality of this resource. The theme of water storage and management is, in fact, fundamental for a territory where rainfall is not abundant, especially in view of the inclusion of activities related to the agriculture and hospitality. Each building has been equipped with a rainwater collecting system that is connected to the grid via a piping system along

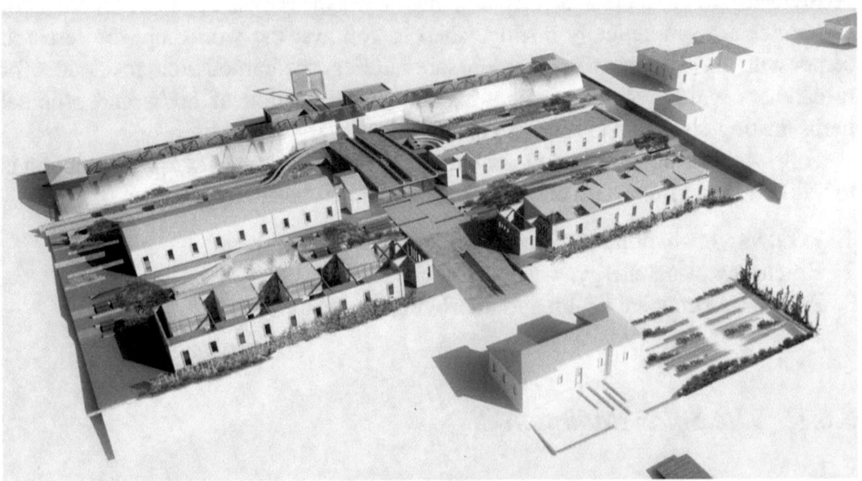

Fig. 8.15 General view of the project. The green promenade, the secret gardens, and the green space laboratories

the main axis, allowing it to be conveyed to the inside of the phytoformulation system, sized according to the actual need. The recovery of the meteoric waters is not enough to satisfy the water demand of the complex and is why, as white waters, so the gray ones in a separate site are naturally purified to go back for irrigation or sanitation. The path that the water makes is well visible: it, on rainy days above all, becomes vitalizing element of the complex. There are many water tanks in the buildings or in the pertinencial spaces not only for aesthetic but also functional purposes: each of them becomes a basin for cooling the surrounding environments on sunny days and a collection tub for the water drainage to the purification system. Once cleaned, the water is collected in a special cistern upstream of the complex placed under the fountain symbolically by the portal, to be redistributed, using the natural slope of the soil, the various buildings and various irrigation systems The mobile shading system, used to provide continuous coverage when needed, has also taken this into account: coverage from the "petals" system in waterproof membranes allows water collection. There is a clear need for rigging of each element's drums and the creation of inspectionable wells at the end of each row.

8.6.3 *The Green*

The study and the design of green have a fundamental role in the project: inspiring from ancient courtyards, secret gardens, and aromas that sparkle in ancient spices tries to dominate the space where it is possible to relax, characterizing the more private ones. The secret gardens will create a deep mix between nature and built, as the island suggests in turn, which has led to the creation of small secluded gardens where residents, students, and researchers could relax in complete peace and silence. In this place more than ever, the chromatic design of the exterior, as well as the interior of the residences, takes on a dominant role: each courtyard has a dominant vegetable essence in order to create a distinct succession of aromas and colors with chromotherapy function. Color is a functional element in the modification of our mood, and that is why every living unit, with its external pertinence area, will stimulate a specific emotional nuance (Fig. 8.16a, b).

8.7 The Solar Green Space and Double-Glass PV System

The solar greenhouse becomes the protagonist of the project. The combination of several functions has allowed the design of an architectural complex where energy-saving technologies (passive solar thermal gain and natural ventilation for passive cooling) are integrated to improve the efficiency and the comfort inside. The architectural integration of photovoltaic system will allow to supply the electricity need, improving the quality of daylighting. The solution of glass-glass polycrystalline photovoltaic panels allows to maximize the power of the solar radiation as well as

Fig. 8.16 (**a**, **b**) The comfort and energy-saving contribution of the courtyards and secret gardens

to implement the shading devices during the summer season. This shading system will increase the quality of inside atmosphere, characterized by the tree pillar forest supporting the roof, recalls the delicate light of the natural wood. The characteristics of solar greenhouse, as an internal microclimate control device, can be summarized in four different phases:

- *Winter day*: solar radiation penetrates inside the green space by heating the floor as a storage mass; no obstacles at the entrance of the solar rays that will heat up the entire classroom. The didactic spaces are aligned behind the south wall to enjoy the benefits of this irradiation and, keeping openings aligned with those of the facade, do not interfere with natural ventilation.
- *Winter night*: the thermal mass of floor, ground, water, and façade exposed to the south allows gradual release of accumulated heat; the raw earth wall side by side with the existing wall guarantees an effective thermal inertia (Fig. 8.17a, b).
- *Summer day*: with temperature rise, the casing opens when the air passes. The grid at the base of the structure allows the passage of fresh currents that, once heated by the inner microclimate, will succeed through the roof windows. When it comes to the shading system, metallic ties have been inserted to interweave white (reflective) sails as necessary; the structure has been inserted internally for convenience and maintenance.
- *Summer night*: during the coolest hours of the day, water and soil will contribute to the internal air circulation, which is effective from the façade-aligned opening system (Figs. 8.18 and 8.19).

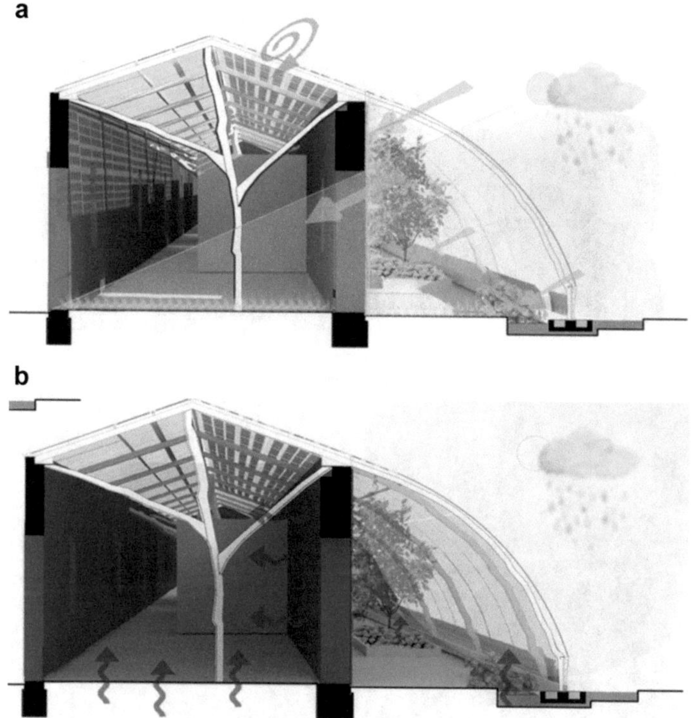

Fig. 8.17 (**a**, **b**) The bioclimatic performance of green space during winter season, day and night

8.8 Pilot Project N.3 Agri Farm and Biodynamic Agriculture Laboratories

The project focuses on the upgrading of the prison building of Campu Perdu as*green building*, dedicated to agri farming and a research lab for biodynamic farmers. There are therefore spaces dedicated to scientific laboratories, reception areas, and teaching and environmental dissemination activities. At the base of the project, even before the architectural concept, there is a long study of the natural environment. The bioclimatic and environmental principles guided the design and the retrofitting action of the historic building organism. The court of the existing building is well suited to exploiting and practicing the principles of natural ventilation. In fact, creating small windows at the bottom and at the top of the opposing sides, you can benefit from every side of the court, the cross ventilation effect created inside the buildings (Figs. 8.20 and 8.21).

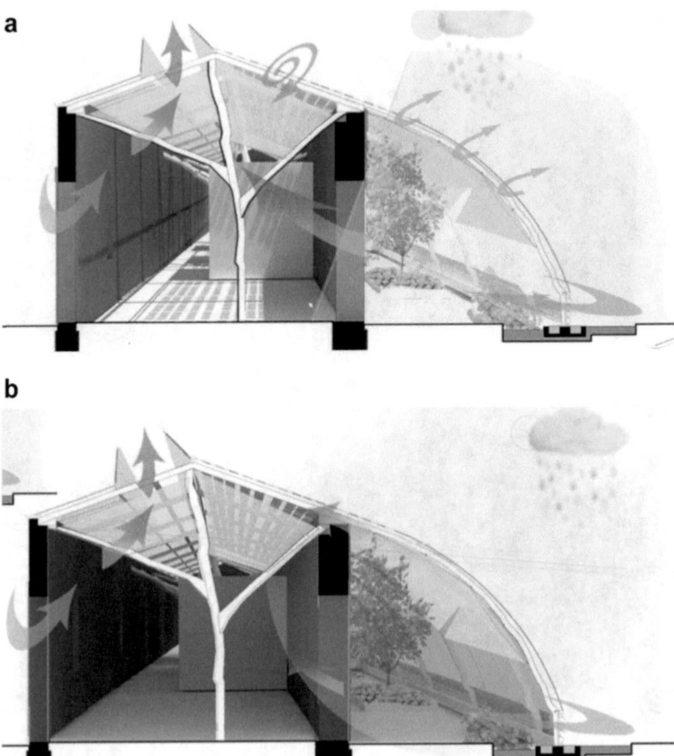

Fig. 8.18 (**a**, **b**) The bioclimatic performance of green space during summer season, day and night

Fig. 8.19 The view of green space dedicated to didactic activities and ancient spices laboratories

Fig. 8.20 The general view of the area of Campu Perdu

Fig. 8.21 The general view of the project of agri farm and biodynamic agriculture laboratories

8.8.1 Natural Ventilation

The microclimate during the summer in the new "square" of Campu Perdu, cooled by the shade of vegetation and the water pool, allows the passive cooling of the air flows. To prevent warm air from leaving the buildings being rebuilt inside, the openings will be protected by rotating shading devices, and the new canopy will control solar radiation improving the comfort of the square. In winter season, thanks to the less lush vegetation, the heat will be accumulated in the floor during the hours of the day and then gradually lowering it overnight. The tensile structures can be adjustable and eventually removable, allowing more solar radiation in winter, contributing to improving solar heating gains and comfort inside and outside.

8.8.2 Renewable Energy

The two main active strategies are the use and integration of solar panels and of a micro-wind. Amorphous solar panels cover the steel canopy. These panels have been integrated on sloping tiles south, southeast, and southwest, allowing them to capture more solar radiation as possible. The micro-wind turbine will be placed at the entrance of the Campu Perdu area, acting as a real sign and landmark. Being low and unattractive, the vertical axis micro-wind device does not present an obstacle or danger to the flying of the birds in the park (Fig. 8.22).

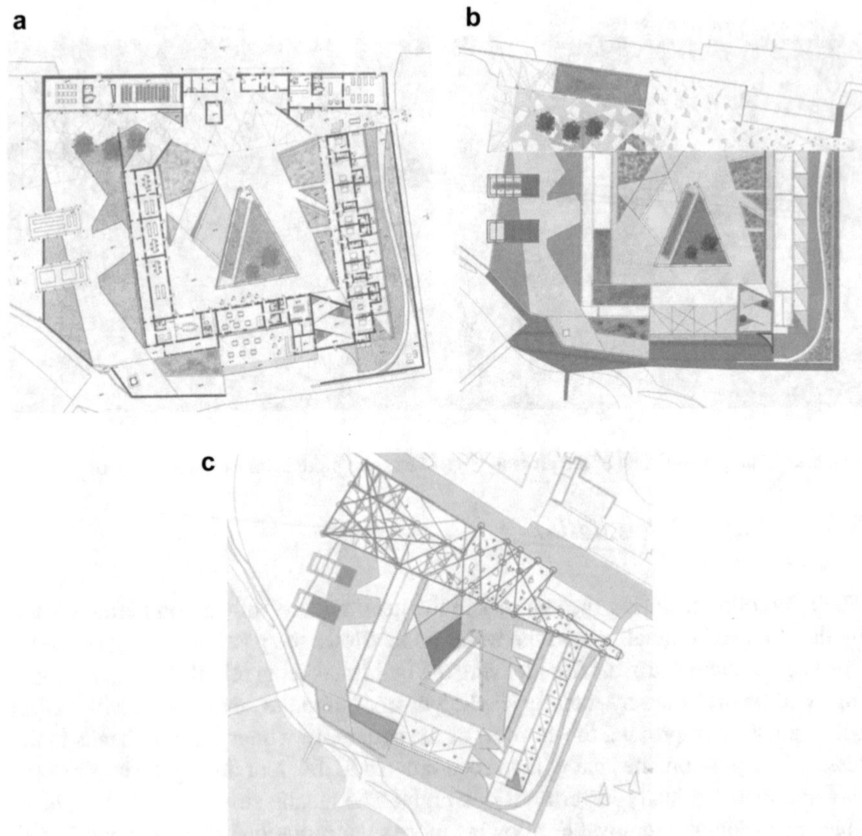

Fig. 8.22 (**a**, **b**, **c**) Plan with functional distribution, green design, and rainwater collection system

8.8.3 The Water

Water is undoubtedly the key element for self-sufficiency on an island. Water in summer season is scarce, so a careful design study is carried out to limit waste and to find new uses for rainwater. Green roofs are designed to collect the rainwater as well as the new canopy. Thanks to appropriate slopes, drainage pipes, drainage, and waterproofing floors, rainwater is conveyed into special storage downstream of the building housing plants for phytotherapy. Once purified, water can be used for irrigation purposes. Flower beds, green frames, and green walls that shield and divide spaces are also repositories and storage of rainwater (Figs. 8.23, 8.24, and 8.25).

8.9 Pilot Project N.4 Fornelli Cultural Hub

The prison of Fornelli and the relevant structures represent a complex of its own. The character and the large size of the structure present an opportunity for the creation of a multipurpose center with a full offer for visitors, a port fate of the island, within which there will be different functions, both public and private. It involves the construction of living labs for research activities, accommodation/suites for researchers, and half-open space dedicated to events and theatrical performances. In addition, part of the prison will be maintained and proposed as part of a museum. Finally, it provides for the creation of an additional restaurant dedicated to the tasting of local products (Figs. 8.26, 8.27, 8.28, 8.29, and 8.30).

Fig. 8.23 General view of the agri farm and biodynamic agriculture laboratories

Fig. 8.24 The restaurant terrace

Fig. 8.25 View of the square under the canopy

8.10 Pilot Project N.5 BioLab and Sailing Center of Cala Reale

Thanks to the low and smooth seabed, as well as to the wonderful environment resources, the project aims to revitalize the old structure of Lazzaretto in Cala Reale as a BioLab and sailing center, more dedicated to disabled people. The natural shape of the gulf and exposure of the area to prevalent winds and good solar radiation and the favorable climatic condition allowed in the past the cultivation of the vineyard and a lush flora and fauna. The project aims to create a new BioLab Center to study the characteristic of local flora and marine fauna, as well as to introduce the disabled people to the practice of sailing, snorkeling, and diving. The environmental conscious design focuses on the implementation of a green building, integrating

Fig. 8.26 View of the prison of Fornelli from the ancient castle

Fig. 8.27 View of the main court of Fornelli with a scenario of canopy artistic installation

bioclimatic strategy to improve indoor comfort, especially in the accommodation suite areas and restaurant. More attention is placed on disabled people hospitality, taking care to implement paths without architectural barriers, as well as fostering emotional didactic areas and sense-able experience of diving activities (Fig. 8.31).

With new technologies constantly being developed to complement current practices in creating greener structures, the benefits of green building can range from environmental to economic to social. By adopting greener practices, we can take maximum advantage of environmental and economic performance, implementing the environmental awareness of the tourists. The solutions of green construction in this pilot project will provide most significant benefits:

- Environmental benefits: reduce wastage of water, conserve natural resources, improve air and water quality, and protect biodiversity and ecosystems

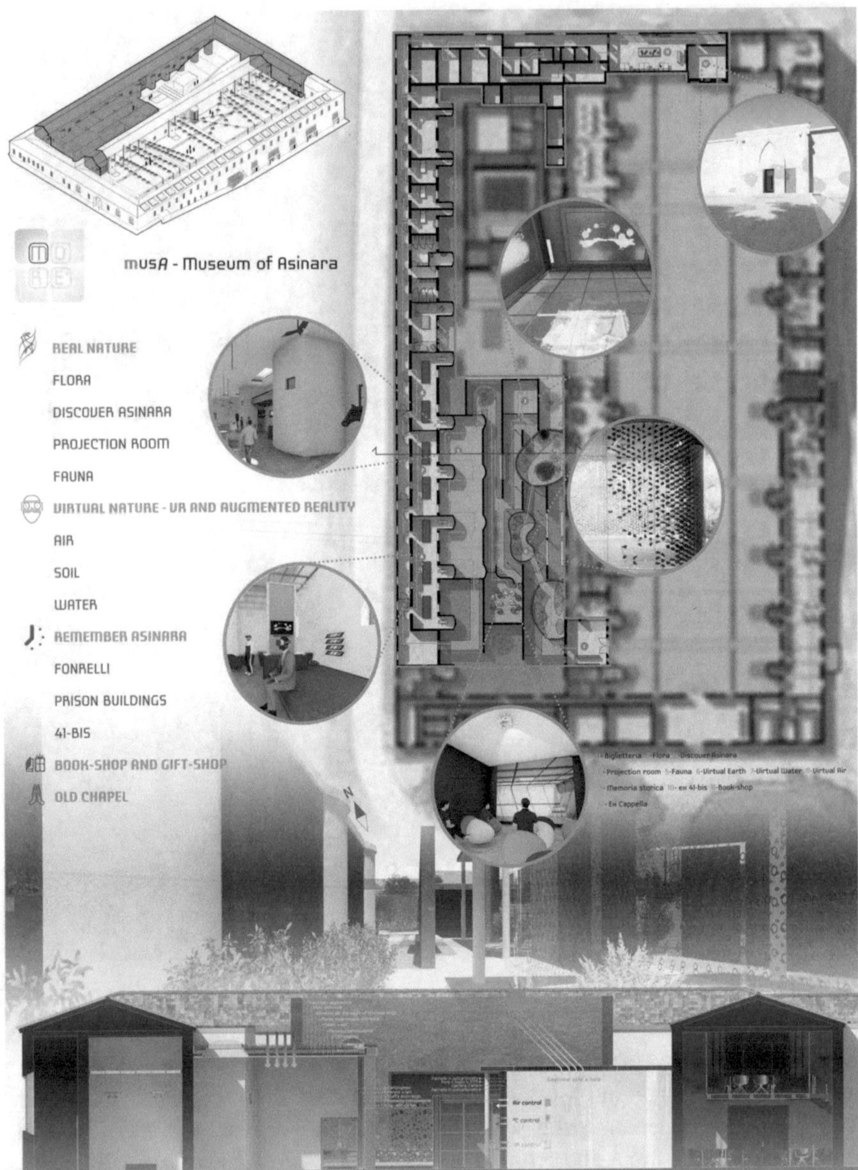

Fig. 8.28 Summary of concept design of Fornelli Cultural Hub: the activities distribution, the exposition roadmap. Bioclimatic strategies to optimize environmental comfort design, improving quality of daylighting, and integrating natural resource (green and water) and renewables

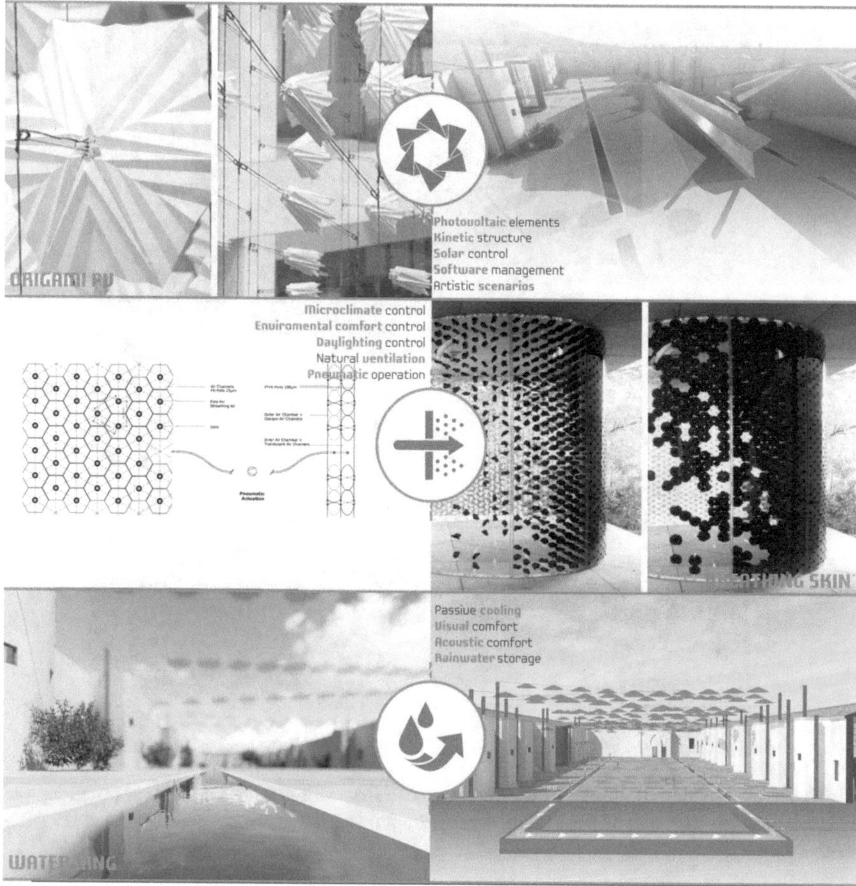

Fig. 8.29 Tables of innovative technologies to improve the quality of microclimate and comfort of external spaces

- Economic benefits: reduce operating costs, improve occupant productivity, and create market for green product and services
- Social benefits: improve quality of life for disabled people, minimize strain on local infrastructure, and improve occupant health and comfort (Figs. 8.32 and 8.33)

The concept design focuses on passive cooling strategies and natural ventilation. Thanks to the quality of the surrounding environment and the proximity of the seaside, the path of the central axis has been enhanced with a floating pier on the sea, useful for disabled people. The full immersion in nature allows this building to become unique and extraordinary. Protected by external existing walls, inside the buildings, there are many covered or open courtyards related to the flora and fauna research activities. It is possible to interact with the external landscape with different access and points of view. The double layers of roof provide a positive natural ventilation, dissipating the overheating during summer season and maintaining high

Fig. 8.30 (**a**, **b**) Views of the exposition courtyards and exposition path. The large use of green and water, integrated with natural ventilation strategies, allows to improve passive cooling in the summer season

Fig. 8.31 General view of the area of Cala Reale and the integration of the new project

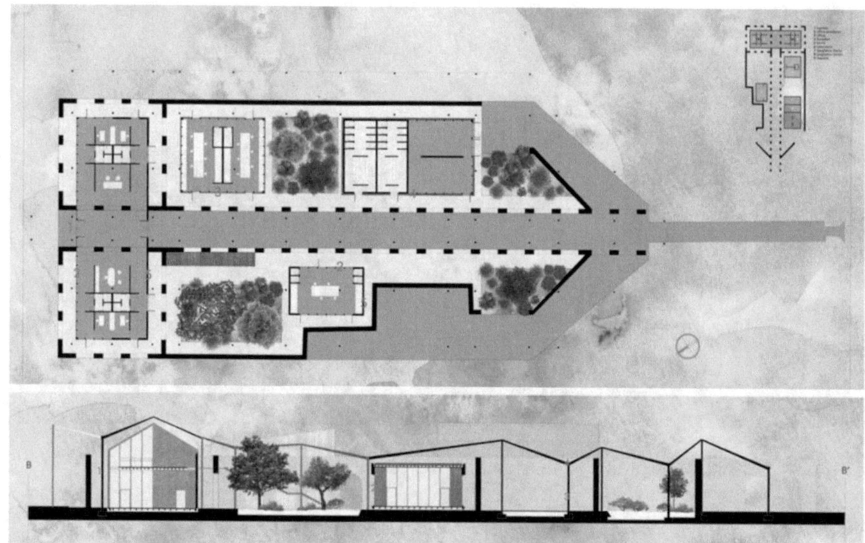

Figs. 8.32 and 8.33 Project plan with main axis to the sea, courtyards, and workshop. The spaces of research present a double ventilated and massive skin

levels of interior comfort. The use of green and water in the inner courtyards helps to create a very comfortable microclimate and wellness experience (Fig. 8.34).

8.11 Pilot Project N.6 Bio-winery and Test Center of Trabuccatu

The project aims to regenerate the old winery and the annexed existing buildings, in use at the prison of Trabuccato, implementing the volumes of services and accommodation in order to answer the new tourist sustainable approach of the Asinara island. The environmental conscious design focuses on the reuse of the old wine cellar function to create an architectural organism that is in harmony with the natural environment and which creates attraction from the tourist's point of view, lovers of wine, and winery researchers and experts. So that it can adopt the cellar system as a trademark of the natural park's publicity, with the aim of expanding receptivity even in the less affluent times. The project therefore proposes as a central theme the production of wine, explaining the production phases as well as the opportunity for the tourists to experiment and test a full immersion experience in a natural dynamic contest. The goal is to design a productive and receptive structure that integrates the theme of eco-sustainability through architectural and technological solutions in symbiosis with the natural environment of the park (Fig. 8.35).

Fig. 8.34 General view of the building and landscape integration

Fig. 8.35 General view of the Trabuccatu tower, existing winery building, and landscape integration

Fig. 8.36 Concept design of the bio-winery as mimetic green architecture

Four elements drive the design concept of the bio-winery green building:

1. *Hypogeum building* as a boundary, as well as a connection element and a convergence axis between land and sea, where the wall represents the border of the natural park and the surrounding environment. Wall, as a border, will create the place where

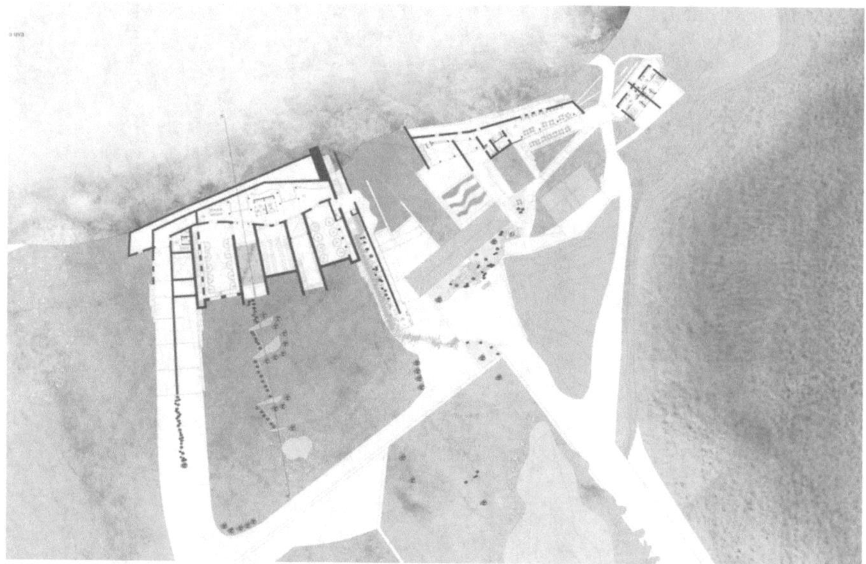

Fig. 8.37 Project plan of hypogeum: mimetic building

the private space dedicated to the memory and the meditation, opens to the world, the fluid boundary that stimulates contact and interaction between interior and exterior. Wall, as a seam, will connect different territories and will allow one to contaminate each other: the contact that triggers human action and makes people meet (Fig. 8.36).

2. *Mimetic architecture*. The idea is to create an architecture that extends the plot of the landscape emphasizing the theme of dry walls (as boundaries) that define the rows of vineyard. The visitor has the first approach, outside, an imperceptible feel of the volumes, discovering afterward the internal space, walking on the green roof, and looking at the tower and the sea, with a full immersion in natural landscape (Fig. 8.37).
3. *The use of natural materials* such as rammed earth and green for landscape design. The bearing walls in fact design the landscape and define the hill according to the level curves. The materials blend perfectly with the colors of the surrounding nature that remains the wealth of the place.
4. *Energy self-sufficiency* and the sustainable use of natural resources: water, green, solar radiation, and natural ventilation (Figs. 8.38 and 8.39).

Fig. 8.38 View of the tower

Fig. 8.39 View of the landscape from testing room

References

AD-Nett, *The European Anaerobic Digestion Network*. http://www.adnett.org [5]

ARUP. (2014). *Cities Alive – Rethinking green infrastructures,* Foresight. [4]

Manuale Alcotra. (2013, Luglio). *La creazione di Living Lab transfrontalieri.* Torino. http://www.alcotra-innovazione.eu [3]

Parco Nazionale dell'Asinara. *Piano del Parco, Relazione Genarale*, resp. Vanni Maciocco. http://www.parcoasinara.org [2]

Trombadore, A. (2015). *Mediterranean smart cities*. Firenze: Edizioni Altralinea. [1]

Chapter 9
24 Bioclimatic Dwellings for the Island of Tenerife: 20 Years Later

Judit Lopez-Besora and Helena Coch Roura

In March 1995, the ITER (*Instituto Tecnológico y de Energías Renovables)*, along with the Excellent Island Government of Tenerife (*Cabildo Insular de Tenerife*), launched an international competition for *25 bioclimatic dwellings to be built in Tenerife*, Canary Islands, homologated by the International Union of Architects (UIA). The International Tender counted on the College of Architects of the Canary Islands to organize a call for preliminary projects of 25 one-family dwellings located in an area within the Wind Park of Tenerife, near the sea, at the south-east of the island (Fig. 9.1). The inception of the competition can be traced back to 1995, in the Summit of the Earth in Rio de Janeiro (1992). One of the key points of this meeting was to make a statement concerning the protection of the environment (General Assembly of the United Nations 2012). Consequently, the projects in Tenerife were to be designed following bioclimatic principles adapted to the climatic conditions (López de Asiaín 2001) of the island's seaside environment. Moreover, the integration of recycled and recyclable materials was encouraged, as well as the use of renewable energy systems.

The competition awakened a great interest around the world resulting in 397 proposals of teams from 38 different countries. The presented proposals displayed a wide range of solutions. The selection committee granted four awards: one first prize, one second prize and two third prizes (Fig. 9.2) (Cendagorta-Galarza and Galván 1996). In addition, other 21 teams were selected and commissioned by the *Cabildo de Tenerife* for the execution of their projects.

J. Lopez-Besora (✉)
Architecture & Energy, School of Architecture, Universitat Politècnica de Catalunya, Barcelona, Spain
e-mail: judit.lopez.besora@upc.edu

H. Coch Roura
Architecture & Energy, School of Architecture, Universitat Politècnica de Catalunya, Barcelona, Spain

Department of Architectural Technology, School of Architecture, UPC, Barcelona, Spain
e-mail: helena.coch@upc.edu

A. Sayigh (ed.), *Seaside Building Design: Principles and Practice*,
Innovative Renewable Energy, https://doi.org/10.1007/978-3-319-67949-5_9

Fig. 9.1 General view of the 24 bioclimatic dwellings location (Source: ITER)

The construction of the 24 dwellings started in a plot next to the ITER headquarters in 2007 and the houses were finally inaugurated in 2010. From that moment on, a monitoring program was implemented to test their long-term performance.

9.1 Introduction

Twenty years later, there is enough collected data to carry out a consistent analysis to assess the implemented solutions. Moreover, the evolution of the bioclimatic architecture parameters through case studies, measurements and literature reviews during these years provides a useful perspective to assess these first attempts of materializing the concept of a bioclimatic project in Tenerife.

9.2 Tenerife and the Canary Islands

Tenerife is one of the main islands in surface and population of the Canarian archipelago, a volcanic group of islands close to the Atlantic African Coast (Fig. 9.3). The first inhabitants of the islands are believed to come from the North African *bereberes*. There is evidence that Carthaginian and Roman citizens established a relation with the local population. Tenerife Island was given the name "Guanche" during the Roman Empire. Current inhabitants of the island are given the same name today. In 1496, all the insular territory became part of the Spanish crown, and ever since, its development has been influenced by the European culture and society.

However, its most important characteristic is its island condition, which means being isolated from other cultures or sources of material and information and surrounded by a vast mass of water that greatly affects its climatic environment.

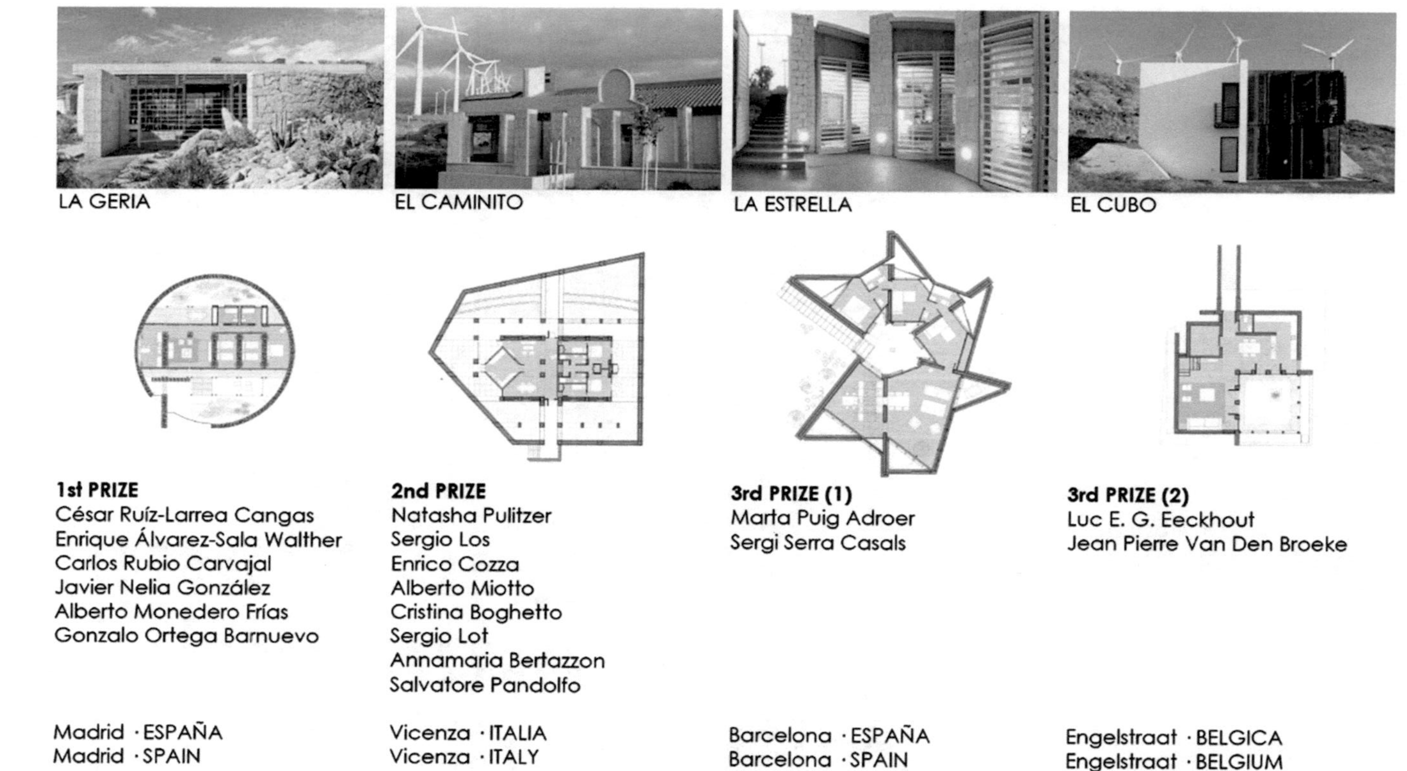

Fig. 9.2 The four awarded projects

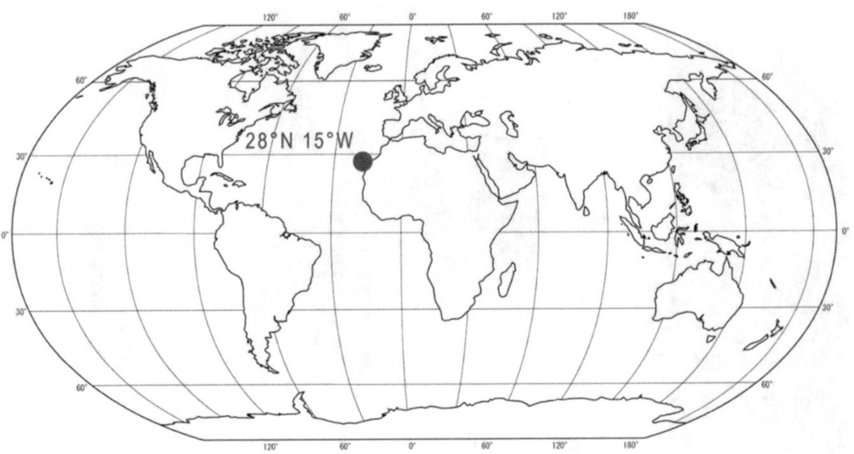

Fig. 9.3 Canary Islands location

9.2.1 Climatic and Geographical Conditions of Tenerife

The *climate* in Tenerife can be considered as semiarid dry-hot (BSh), according to Köppen climate classification (Köppen 1884). It is characterized by scarce rainfall, many hours of sunshine and constant winds (Martin Vide and Olcina Cantos 2001).

Great extents of water mass reduce thermal oscillation (Olgyay 1963). Consequently, *proximity to the sea* generally produces climates with milder temperatures than continental locations. The Canary archipelago, being a group of islands situated in the Atlantic Ocean, has a climate greatly affected by the sea regulatory effect. Therefore, the island's temperature swing is slight along the year. This climatic characteristic, very distinctive of seaside locations, is a central feature of the Tenerife Island.

The mean annual temperature is over 18 °C, quite constant along the year (Table 9.1). The seasonal variation is small, from 18–19 °C in winter to 23–24 °C in summer, which constitutes an even distribution of temperatures along the year (annual variation 4–6 °C). The relative humidity is around 71–72% average most of the year. Therefore, the climatic condition of Canary Islands in terms of temperature can be considered to be within the comfort zone most of the year.

The constant temperature at comfortable values is in close relation with the high amount of sunshine hours and clear-sky condition days in the island. The average of sunshine hours is about 55–60%, as the combination of clear-sky condition days and partially clouded days take up most of the year. The latitude of Tenerife ranges from 28°N to 28.6°N. In this position, the sun's incidence is quite vertical (85° in summer, 38° in winter, at noon), and the daily values of solar irradiation in a horizontal surface can reach up to 8.16 kWh/m^2 (Sancho et al. 2005).

Another characteristic of this climate is the continuous *wind* flow coming from the north-east, called *Alisio* (trade wind). The presence of this wind is seasonal, with

Table 9.1 Climatic data in Tenerife (period 1981–1992) (Source: Cendagorta-Galarza and Galván 1996)

Month	Average temperature (°C)	Average wind speed (m/s)	Hours of sunshine	Clear-sky condition days	Low cloudiness days
Jan	18.42	7.55	191.57	10.33	19.33
Feb	18.36	7.59	192.59	8.50	17.17
Mar	19.41	7.97	210.42	10.17	18.33
Apr	19.24	7.59	199.23	6.50	21.17
May	20.16	7.17	233.13	4.67	23.83
Jun	21.91	7.43	237.08	7.83	21.00
Jul	23.93	8.19	271.26	18.17	11.83
Aug	24.89	8.03	256.63	15.50	15.00
Sep	24.75	7.31	191.18	8.50	20.17
Oct	23.40	7.06	199.49	3.33	24.00
Nov	21.53	7.13	185.61	5.83	21.33
Dec	19.50	7.01	190.82	4.67	24.00
Tot	21.28	7.50	2599.31	104.00	237.17

Fig. 9.4 Typical landscapes of the north side of Tenerife (Anaga) and the south side (Cañadas del Teide)

special frequency during summer (Dorta 1993). A gust of wind can reach more than 30 m/s, although the average wind speed ranges between 7 and 8 m/s. The main direction of trade winds in this place all year long is from the north-east to south-west.

One of the main characteristics of Tenerife is the steep *topography*, which varies in a few kilometres from sea level to 3718 m over sea level on top of Mount Teide volcano. This topographic variation highly influences the local climate, a fact that is visible when finding opposed conditions in the northern and southern slopes of the mountains, being arid at the south side and more humid at the north.

The South of Tenerife constitutes an arid area with less humidity and rainfall. As a result, the vegetation in this area is composed only by bushes and other types of low plants which means a scarceness of lush vegetation (Fig. 9.4).

All in all, the meteorological data in Tenerife present average conditions with very regular values, often guaranteeing human comfort needs. The proximity to the

Fig. 9.5 Human settlements in the island of Tenerife are adapted to climate and the environment characteristics

sea is a macroclimatic factor which acts as a climate regulator, while other factors such as latitude and altitude are also determining (Fig. 9.5). Consequently, as the weather conditions tend to be very favourable along the year, it provides an easy framework to develop a bioclimatic project.

9.2.2 Traditional Architecture in Tenerife

Considering the benevolent climate of the islands, architecture would not be essentially intended as a climate shelter, as in the case of less temperate climates, but as a place designed to satisfy other needs such as security or privacy. Consequently, the challenge of architecture in this place is to guarantee and extend in the indoor spaces the already comfortable conditions of the outside (Cornoldi and Los 1982; Neila-González 2004).

Traditional architecture in Tenerife responds to these environmental conditions with different strategies. However, as a great part of urban developments in Tenerife and Canary Islands were influenced by colonial architecture, it is not easy to find remains of vernacular location-based architecture (Flores 1977). Most urban centres relate to a foreign style, which also characterizes Latin American cities that respond to the political situation rather than climatic factors. Nevertheless, some remains of popular architecture can still be found in the island, mainly in rural areas.

The first aspect to take into account when discussing architecture as a response to climate in this location is the *control of solar radiation* (Mazria 1979; Olgyay and Olgyay 1976; Wright 1983), which can be critical during the warm season. If not managed properly, an excess of solar gain could result in an increase of indoor temperature. Consequently, architectural designs that could facilitate the greenhouse effect are generally avoided in the island's popular architecture. However, the solar radiation contribution is desirable during certain periods of the cold season, which is why filters and screens are broadly used in order to guarantee solar protection and also offer the possibility to allow solar gains in certain periods.

The second characteristic of this climate is the need of *wind management*. The presence of wind in the island has positive and negative consequences and, therefore, architecture is in need of a double-edged strategy when facing this factor

Fig. 9.6 Vineyard plantations in Canary Islands (Author: Nina Amat)

(Givoni 1994). On one side, in the warm period, wind is key to guarantee comfort conditions through wind-chill reduction of perceived temperature (Santamouris and Asimakopolous 1996; Allard 1998). On the other side, wind excess might bring practical discomfort and the impossibility to perform certain activities. This factor can be perceived when local traditional means of growing vineyards are observed (Fig. 9.6). The seeds are planted in small depressed areas, protected from the wind by a middle moon shaped enclose made of terrain. Moreover, in cool periods, protection from the wind is necessary to guarantee comfortable perceived temperature. Half of the year, *Alisio* blows with constant intensity and in the same direction. This regularity eases the design of a protection to avoid the constant wind gust. Being the wind direction constant, the barrier can be permanent, located in the north-east side. Vineyard plantations are a clear expression of the need for this permanent barrier.

As a response to local climate, the location, shape and elements of buildings have adapted for years to the main climatic features of the island, which are the sun and wind. For a long period while mechanical equipment had not been integrated yet, the main strategy was to give shelter from radiation and the wind. This condition was approached through two main architectural elements: courtyards and facades' elements intended to filter the connection between indoor and outdoor conditions, such as semi-covered balconies or small openings in exposed orientations (Feduchi 1978; Luna and Lucas 2007) (Fig. 9.7). Both strategies offer a protection from wind, functioning as a barrier if they are strategically located. At the same time, they are also effective to filter or manage the incidence of solar radiation in floors and walls, in the case of courtyards, and over the openings in the case of windows and balconies.

The use of *courtyards*, usually facing south and south-east, allows the management of ventilation into the building, both creating outdoor spaces protected from the strong wind and offering a smoother transition from indoors and outdoors spaces. In addition, the excess of relative humidity in certain periods of the year or day can be reduced with the crossed ventilation provided by the presence of the courtyard. All in all, traditional courtyards clearly respond to the architecture needs in Tenerife by providing a space that guarantees security and privacy with improved outdoors climate conditions through wind and radiation control.

In the same line, the design of façades also responds to the protection of sun and wind. Traditional architecture is characterized by the presence of very distinctive *balconies*. Although currently being appreciated for their popular characteristic

Fig. 9.7 Traditional courtyard in the Montañés house and balcony in the Cabrera Pinto Canarias Institute façade in San Cristobal de la Laguna

style, they are also a consequence of the need for dealing with solar radiation and wind through filters, barriers and connectors (Serra Florensa and Coch 2001). Their structure, covering most of the balconied area with different types of openings and adaptive mechanisms, provides sun protection, light access, night insulation, wind protection and ventilation. Few openings located on south facades and an extremely reduced number of them at north and north-east facades offer a good response to the needs of this seaside place.

In order to achieve these architectural strategies, two main materials are used by traditional architecture in Tenerife. Local volcanic stone confers heaviness to walls and the building envelope guaranteeing the security and wind protection needed. This material is often used as the primary massive element of construction, creating the basic design lines and aiming to protect from permanent non-wanted inputs, such as the Alisio wind. On the other side, wood is generally used for horizontal structures and other attached elements, such as balconies and other types of canopies and galleries. This lighter material offers flexibility to the ensemble by providing filtering devices adaptable to different needs throughout the day and year.

Although *vernacular architecture* in Tenerife has been pushed aside by the introduction of Spanish colonial developments, some remaining buildings or elements are still able to show a glimpse of traditional strategies that answer to environmental conditions.

9.3 Bioclimatic Architecture

The "bioclimatic architecture" concept is a term used since the second half of the twentieth century. Vernacular architecture was built for centuries under the principles of local climate adaptation and the availability of materials in the immediate surroundings (Olgyay 1963; Serra Florensa 1989). However, these principles started to disappear once mechanical equipment aiming to guarantee the user's comfort was incorporated in architecture. Through the first and second oil crisis, a concern for overuse of fossil fuels raised, together with the increasing lack of connection between architecture and environment, which led to the irruption of a new way to design buildings called *bioclimatic architecture*. In recent years, the increasing concern about global warming, carbon emissions and energy consumption (Houghton 2004; Flavin 1989) keeps the need for research on this field up to date as well as the development and implantation of renewable energies.

9.3.1 Bioclimatic Architecture in Tenerife

In the island of Tenerife, in addition to the concern about global warming and protection of the environment, two other factors determine the interest for bioclimatic architecture and renewable energies' research: the isolated condition of the place and the existence of a dense community with increasing specific tourism.

The archipelago's isolation from the mainland and the rest of the country results in a need for self-sufficiency in terms of energy and materials. For this reason, the research on renewable energies is key to the autonomous functioning of the island, while the seaside condition of the place configures specific demands in terms of bioclimatic architecture and energy use.

The second key factor is the high population, increased by tourism in certain periods of the year. According to data from ISTAC (*Instituto Canario de Estadística*[1]), Tenerife's population in 2016 was of 891,111 inhabitants, while in the same year 6,595,560 people were counted to have visited the island. The volume of temporary population implies a higher demand of energy, which aggravates the dependence on the territory. Moreover, the water supply and its quality is another main concern for the inhabitants of the island.

As a response to this situation, political stakeholders established in 1990 the *Instituto Tecnológico y de Energías Renovables* (ITER[2]), whose mission is the achievement of a high degree in research and development of alternative energy sources in order to diminish the dependence of the island. Nowadays, this institution is in charge of several research projects regarding renewable energies, one of which is the management and tracking of the bioclimatic dwellings performance built as a result of the competition.

[1] http://www.gobiernodecanarias.org/istac/

[2] http://www.iter.es/

Fig. 9.8 Site plan and list of house names

9.4 Results of the Competition

9.4.1 The Construction of the 24 Bioclimatic Houses

The design of the 24 dwellings in the island of Tenerife provided an opportunity to test the bioclimatic principles in the Canary Islands, including the vision of international teams and their particular analysis of the architectural needs. A first observation of the ensemble underlined the variety of appearances, materials and shapes. This diversity could have been a result of the multiple background and tastes of the design team. The construction of the 24 dwellings started in 2007 and was finally inaugurated in 2010 (Fig. 9.8).

However, both the objectives and strategies presented by the participants have a common ground. The result is an architectural collection of 24 dwellings with similar but not identical solutions for bioclimatic construction in the island of Tenerife.

9.4.2 Bioclimatic Strategies

An interesting observation to mention is the recognition of a unifying tendency related to the main bioclimatic strategies applied in this climate beyond the differences among the architectural design of the projects. The common strategies can be summarized as following: direct solar gain, ventilation, solar radiation protection and wind protection, in addition to the use of solar active systems (photovoltaics and solar thermal). In Fig. 9.9, a synopsis of the strategies applied in the projects is presented in the following table.

	SOLAR GAIN	BURIAL DEGREE	VENTILATION SYSTEM	SOLAR PROTECTION SYSTEM	PHOTOVOLTAIC PANELS	SOLAR THERMAL PANELS
ALISIO	●●○				●●○	●○○
ARCILLA	●○○				●●●	●●○
BERNEGAL	●●○				●●○	●●○
BERNOUILLI	●○○				●○○	●○○
BOVEDAS	●●○				●●○	●●●
CAMINITO	●○○				●●○	●●●
CANGREJO	●●●				●●○	●●○
COMPACTA	●○○				●●○	●○○
CUBO	●●○				●○○	●●●
DISPOSITIVO	●●●				●●○	●○○
DUNA	●●●				●○○	●○○
ESCUDO	●●●				●●○	●○○
ESTRELLA	●●●				●○○	●○○
GAVION	●●●				●●●	●○○
GEODA	●●●				●○○	●●○
GERIA	●●●				●●○	●●○
MURO	●●●				●○○	●●○
NOCHE Y DIA	●○○				●○○	●●●
PATIO	●●○				●○○	●○○
PUEBLO	●●○				●●○	●●○
RELIGA	●●●				●●●	●○○
RIO	●●○				●○○	●●○
TEA	●●●				●●○	●○○
VELA	●●●				●●●	●●○

Fig. 9.9 Chart with the list of dwellings and the implemented bioclimatic strategies

The strategies described in Fig. 9.9 are classified as *passive systems* (columns 1–4) and *active systems* (columns 5–6) (Givoni 1969). In general terms, passive systems can be sorted in: solar gaining, thermal inertia, air control systems and solar protection (Serra Florensa and Coch 2001). In the climate of Canary Islands, we find that inertia systems are not appropriate due to the little variation in temperatures along the year, while ventilation and solar protection combined with solar gain are the most suitable strategies. In addition, wind protection is also a central strategy needed in this climate.

Fig. 9.10 Houses *La Estrella and La Compacta,* partially buried

Wind protection has been implemented in the projects using different methods that can be divided in three categories: by partially burying the volume, as in the case of *La Estrella* or *La Compacta* (Fig. 9.10), by using high walls to protect both the building and the courtyard as seen in *La Geria* or *La Duna* or through a lack of openings on the north and north-east facades, a resource found in most of the buildings.

On the other hand, *Alisio* is not only considered as a negative phenomenon to be protected from, but sometimes, mainly in hot and humid conditions, it is used as a ventilation system. The dwellings are located in a plot next to the seaside, implying that relative humidity can be critical in certain periods, increasing the heat index and, therefore, the perceived temperature. Those systems mainly consist in crossed and forced ventilation. Some of the dwellings like *La Duna* introduce wind towers to take advantage of the regular wind flows coming from the north-east (Fig. 9.11), producing a negative pressure which contributes to the removal of the warm air inside. Other systems use air cavities and static chimney cowls for the same purpose (*La Geoda*, *Las Bóvedas*, for example), as shown in Fig. 9.11. These systems, along with crossed ventilation, constitute the main ventilation strategies implemented in the projects. Although the most used system is crossed ventilation, usually between opposite (north-south) façades, many projects combine different ventilation systems to cool down the indoors thermal conditions.

Solar gain is fundamental in the design of a bioclimatic project and is not only used for passive systems but is also key to implement active ones, such as PV cells or thermal panels. The houses are covered with different degrees of *solar active systems* related to thermal energy gains and photovoltaic generation power (Fig. 9.12), integrated in the main architectural surfaces, especially in roofs (*El Muro* or *La Duna)*. The implementation of solar systems in horizontal or almost horizontal planes is possible due to the latitude of Tenerife (28° N) and the consequent sun position (85° altitude in summer). In most of the cases, the solutions implemented in the dwellings are a guarantee for self-sufficiency.

Fig. 9.11 Wind towers in the north façade of *La Duna* house; static chimney cowls for ventilation in *Las Bóvedas* and *La Geoda* houses

Fig. 9.12 Solar active systems in the roof of *El Muro* and *La Duna* houses

The climate of Tenerife sets *solar protection* as a priority before solar gain. There is a wide array of solar protection systems, which have been classified in Fig. 9.9 in three types: horizontal protection (porches, eaves, etc.), external vertical protection (blinds, shutters, lattice, etc.) and internal vertical protection (curtains, screens, etc.). Each of them has a different degree of protection, acting as filters or barriers in all cases (Fig. 9.13).

The use of solar protection has been implemented in all houses with different solutions, permanent or mobile depending on the characteristics of the system. In the case of passive solar systems, the challenge is to find the balance between gain and sun protection, as a radiation excess would produce an indoor temperature rise, critical at some moments of the day and certain periods of the year. However, when the external temperature decreases under the comfort level, solar gain becomes necessary.

Fig. 9.13 Solar protections in *La Compacta* and *La Estrella* houses

This sample of *seaside bioclimatic houses* is an interesting experimentation field that allows a monitoring of the implemented systems' performance. The opportunity provided by the existence of 24 samples of house sharing a common objective and strategies allows for an assessment and comparison of indoor conditions to be performed.

9.5 La Geria: 20 Years Later

Since the construction of the dwellings in 2007 and their inauguration in 2010, a complete tracking program has been carried out in each house. In addition, the dwellings have hosted a wide range of users through a renting system managed by ITER that allows visitors to inhabit the house during different time periods. The ITER institution has collected several data from specific physical parameters in the buildings which, along with the feedback taken from visitors who have been accommodated there, are an excellent source of information. Moreover, the institution offers the option of reviewing the values of the parameters of each house in real time on the living room TV, increasing the pedagogical purpose of the initiative.

The interior of the houses has been provided with sensors to measure different parameters: relative humidity, air speed, air temperature at different heights, superficial temperature and ceiling cavities temperature (Fig. 9.14). Electricity metres are also available in the houses. In addition, a meteorological station that records weather data is located in the same plot. All this information is recorded minute by minute and stored for further analysis.

The meteorological parameters in the location where the homes were built are constantly measured by the ITER institution. Displayed in Table 9.2 are the year 2013 corresponding values. Even though the wind speed in 2013 was lower than the average in the 1981–1992 period, the temperatures are in between 18 and 24 °C, which coincide with the values in Table 9.1. A new value in this table is the relative humidity range which varies from 68 to 77%, a constant along the year.

Fig. 9.14 Temperature and humidity sensors in the ceiling of the houses

Table 9.2 Meteorological data taken in the surroundings of ITER institution (2013)

Month	Radiation (MJ/m^3 day)	Average ambient temperature (°C)	Maximum precipitation (mm/month)	Average ambient humidity (%)	Wind speed (m/s)	Predominant wind direction
Jan	13.44	18.40	47.00	69.00	4.94	NE
Feb	17.00	19.20	61.80	74.00	5.99	NE
Mar	17.99	19.10	101.20	71.00	5.71	NE
Apr	19.82	19.40	35.60	71.00	6.18	NE
May	23.91	19.80	6.70	69.00	5.43	NE
Jun	25.40	21.20	1.90	71.00	5.63	NE
Jul	27.05	22.60	1.50	73.00	7.85	NE
Aug	22.85	24.10	1.60	75.00	7.40	NE
Sep	18.67	22.80	19.20	77.00	5.62	NE
Oct	17.62	23.00	53.70	77.00	5.10	NE
Nov	14.20	21.40	212.80	71.00	5.33	NE
Dec	11.35	20.10	83.80	71.00	5.90	NE

9.5.1 *"La Geria", the Winner Project*

This section presents the main results of the monitoring program in one of the houses, *La Geria*, in order to outline the principal recommendations for a bioclimatic project in a seaside dwelling in the island of Tenerife. *La Geria*, designed by the team of the architect Cesar Ruiz-Larrea, was awarded the first prize of the competition and displays some of the discussed bioclimatic strategies.

Fig. 9.15 Solar protection systems in the north courtyard and at south façade in *La Geria* house (Source: ITER)

9.5.2 Description of Bioclimatic Strategies

The building is organized in an east-west axis that generates two main façades with southern and northern orientation. These façades are opened towards two courtyards surrounded by a circular wall that protects the unit from constant winds. Both façades are constituted by a 90% glazed surface, the southern one becoming an important collection of direct solar gain. However, this transparent envelope is protected from solar radiation with different elements (Fig. 9.15). A continuous pergola made of wooden louvers covers a great part of the façade with the objective to protect it from the vertical sun beam. At the same time, there is a 90 cm roof overhang as well as a 40 cm deep set of square shelves for plants that has been placed in front of the kitchen zone, with the purpose of protecting it from solar radiation.

On the other hand, the air exchange relies on *crossed ventilation* created by the two courtyards located on the opposite orientations. This system is reinforced by another air flow that traverses the space: the air intake is placed at floor level, provided by a cavity under the floor connected with the exterior; the extraction is carried out through another cavity under the roof, also connected with the exterior. The connection between both cavities is accomplished by ventilation grilles at floor and roof level inside, which allows for an adjustable ventilation to be managed in the space. The air supply from the exterior is magnified by the action of the wind (Fig. 9.16).

9.5.3 Data Collected Analysis

The measurements of air temperature and relative humidity carried out at the same time intervals enable a comparison of the indoors and outdoors conditions to be made. Contrasting the temperature and relative humidity values in La Geria, as a result of an exhaustive study of an extensive period of time both inside and outside, a first appraisal can be made. On average, the values of air temperature inside the house (Fig. 9.17) range from 4 to 5 °C above the outside temperatures, and this

Fig. 9.16 Different parts of the ventilation system in *La Geria* House

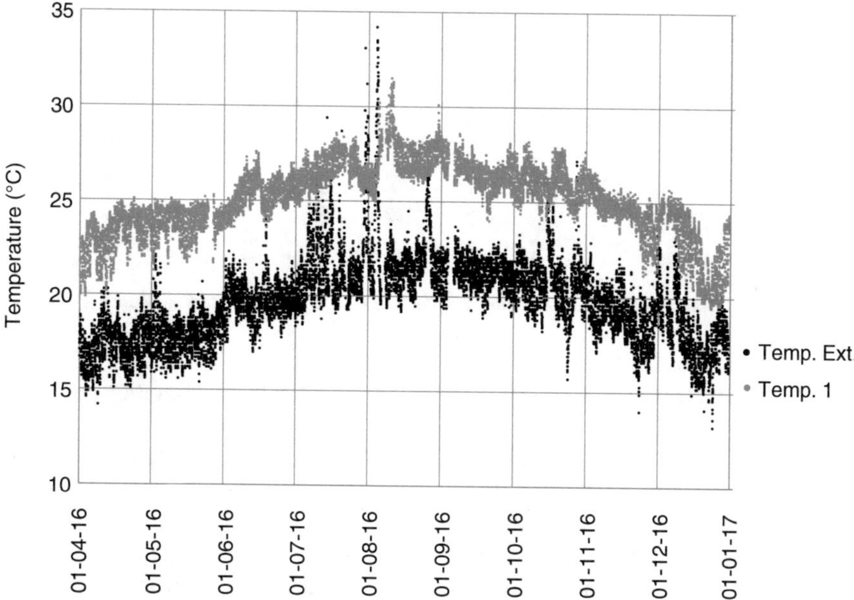

Fig. 9.17 Outdoor and indoor temperature in *La Geria* house in the period March–December 2016

difference is quite constant throughout the year. Considering that the annual variation of temperatures in this climate is around 5 °C, and the average temperatures are 18–24 °C, the temperature inside ranges between 22 and 28 °C. In addition, the thermal fluctuation on the inside is slightly lower than outside. It can be concluded that, in this case, architecture contributes to stabilize the temperature.

Observing the value of relative humidity over a long period of time (Fig. 9.18), the indoor data presents a decrease compared with the outside. The variation reaches about 30% in certain periods, which seems to indicate that ventilation systems are successful in reducing indoor relative humidity. Although the highest outdoor values of relative humidity throughout the year are in summer, the variation along the year is not very significant, and in the inside of the house, this variation is reduced. On average, relative humidity remains around 50% in the interior of the house.

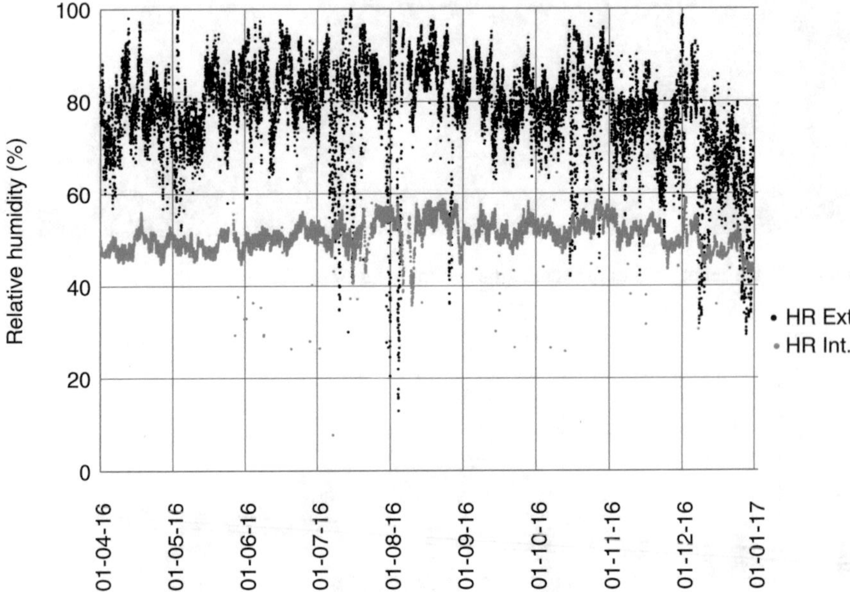

Fig. 9.18 Outdoor and interior relative humidity in La Geria house in the period March–December 2016

However, the wind action produces sudden drops of relative humidity in certain periods.

A period of 1 week in the middle of May 2016 is presented in order to analyse the results more thoroughly (Figs. 9.19 and 9.20).

The different temperature sensors in Fig. 9.19 correspond to air and surface temperature at different positions, hence the variation of values among them. The temperature inside always remains over the values of the exterior, from 5 °C in the coolest space to 8–9 °C in the warmest one. The steadiest temperatures correspond to air temperature, while the most changeable values correspond to superficial temperatures. The contribution of solar protection in this example is noteworthy, because the south façade is protected by different systems, which prevent the house from overheating. The graph also shows the increase of temperature inside and a slight phase difference between the oscillations outside.

In terms of ventilation, if relative humidity inside and outside are compared (Fig. 9.20), it can be noticed that the values inside do not surpass 50%, always staying close to this value. The same parameter fluctuates approximately between 60 and 80% on the outside, which implies a decrease of relative humidity inside around 15–38%. The cause of this reduction can be traced back to ventilation systems, specially crossed ventilation. In addition, the results and the feedback of the building's users show that forced ventilation through floor and roof cavities seems to be an effective system.

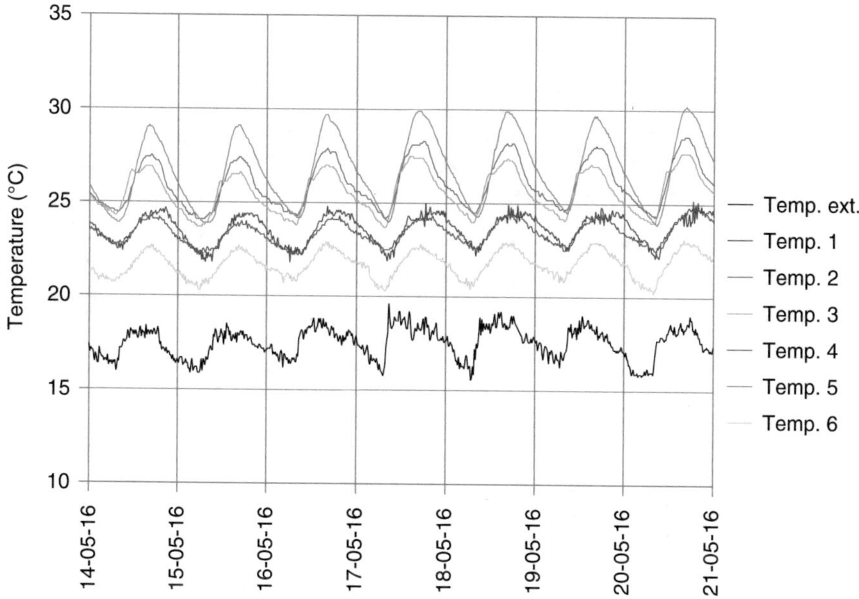

Fig. 9.19 Outdoor and interior temperature in La Geria house in the period 14/05/2016–21/05/2016

9.6 The Evolution of Vernacular Architecture

La Geria is an example of the implementation of bioclimatic strategies in a mild climate, being part of a group of 23 more dwellings that also offer the opportunity to learn from them (Coch 1998). The quantity of data collected makes difficult to show all the results but it can be stated that the other houses of the competition have presented similar results in terms of collected data and user's feedback.

In general, the balance between solar protection and solar gain has shown to offer a good performance in most of the houses, only with overheating peaks in the projects where these strategies have been insufficiently integrated in the design. Similar considerations can be extracted about the performance of ventilation systems. In this case, air movements are more difficult to predict because of the variability of forced ventilation and wind. In general, crossed ventilation is an easy system to implement in a single house, and the volume of exchanged air is considerable.

The existence of this laboratory in continuous evolution offers a great opportunity to learn from the bioclimatic experience in situ on one side and to get feedback from the results and experiences from its users. What we have learnt from this experience is that, although it might seem easy to design comfortable shelters in mild climates, it is important to look at, listen to and touch the place where we build in order to preserve the positive aspects of the place.

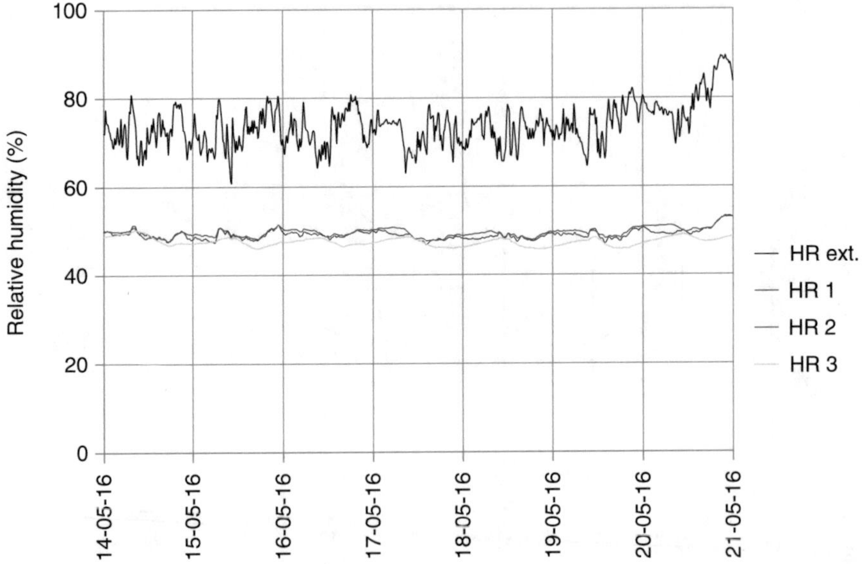

Fig. 9.20 Outdoor and interior relative humidity in La Geria house in the period 14/05/2016–21/05/2016

9.7 Conclusions

All in all, this assessment has offered the opportunity of acquiring an overview of the 24 bioclimatic dwelling strategies and performance. The most successful design options implemented in the houses are related with the presence of wind protection, courtyards and façade filters to regulate the connection with the exterior. These approaches are very similar to traditional architecture main features.

For instance, the circular-shaped wind protector walls of *La Geria* highly resemble the vineyards implantation strategies, while the presence of courtyards is a common feature of popular buildings. Finally, the semi-covered balconies displayed in traditional architecture are a compendium of the different filters present in the 24 bioclimatic dwellings façades.

Nonetheless, the analysed houses also implement active systems that gain energy through solar radiation or wind speed. It can be concluded that the houses success is based on design strategies very similar to traditional architecture of the area, complemented with active energy gaining systems.

References

Allard, F. (1998). *Natural ventilation in buildings: A design handbook*. London: James & James.

Cendagorta-Galarza, M., & Galván, G. (1996). *Twenty-five bioclimatic dwellings for the island of Tenerife*. Tenerife: ITER.

Coch, H. (1998). Bioclimatism in vernacular architecture. *Renewable and Sustainable Energy Reviews, 2*, 67–87.

Cornoldi, A., & Los, S. (1982). *Hábitat y energía*. Barcelona: Gustavo Gili.

Dorta Antequera, P. (1993). *El clima: tipos de tiempo. Geografía de Canarias* (Vol. 1). Las Palmas de Gran Canaria: Prensa Ibérica.

Feduchi, L. (1978). *Itinerarios de arquitectura popular española. 4-Los pueblos blancos*. Barcelona: Editorial Blume.

Flavin, C. (1989). *Slowing global warming: A worldwide strategy*. Washington: Worldwatch Institute.

Flores, C. (1973–1977). *Arquitectura popular española. Parte cuarta: la arquitectura popular en el archipiélago canario*. Madrid: Aguilar.

General Assembly of the United Nations. (2012). The future we want; A/RES/66/288*.

Givoni, B. (1969). *Man, climate and architecture*. Amsterdam: Elsevier.

Givoni, B. (1994). *Passive and low energy cooling of buildings*. New York: Van Nostrand Reinhold.

Houghton, J. T. (2004). *Global warming: The complete briefing*. Cambridge: Cambridge University Press.

López de Asiaín, J. (2001). *Arquitectura, ciudad, medioambiente*. Sevilla: Universidad de Sevilla Secretariado de Publicaciones.

Luna, M., & Lucas, M. (Eds.). (2007). *Arquitectura tradicional y entorno construido*. Murcia: Trenti.

Martin Vide, J., & Olcina Cantos, J. (2001). *Climas y tiempos de España*. Madrid: Alianza.

Mazria, E. (1979). *The passive solar energy book*. Emmaus: Rodale Press.

Neila-González, F. J. (2004). *Arquitectura bioclimática en un entorno sostenible*. Madrid: Munilla-Lería.

Olgyay, V. (1963). *Design with climate: Bioclimatic approach to architectural regionalism*. Princeton: University Press.

Olgyay, A., & Olgyay, V. (1976). *Solar control and shading devices*. Princeton: Princeton University Press.

Sancho, J. M., Riesco, J., Jiménez, C., Sánchez de Cos, M. C., Montero, J., & López, M. (2005). Atlas de radiación solar en España utilizando datos del SAF de Clima de EUMETSAT. http://www.aemet.es/es/serviciosclimaticos/datosclimatologicos/atlas_radiacion_solar

Santamouris, M., & Asimakopolous, D. (1996). *Passive cooling of buildings*. London: James & James.

Serra Florensa, R. (1989). *Clima, lugar y arquitectura*. Madrid: CIEMAT DL.

Serra Florensa, R., & Coch, H. (2001). *Arquitectura y energía natural*. Barcelona: Edicions UPC.

Wright, D. (1983). *Arquitectura solar natural: un texto pasivo*. México: Gustavo Gili.

Chapter 10
Design of Seaside Buildings in China

Marco Sala and Antonella Trombadore

10.1 The Great Potential of Coastal Cities as Tourist Attraction

Seaside towns in China have been considered for century mainly as commercial gate to other countries or as productive base for fishermen and fish industry. Only in recent time, China discovered the great potential of coastal cities as tourist attraction, both from inland and from abroad. Wye Hay town and its peninsula, which for hundreds of kilometers protrude toward the Yellow Sea, have been recently subject of strong building activities to provide hotels, marina, and private houses for the new tourist development of China. Nevertheless, the architectural approach is many times oriented toward the "standard" model of urbanism and architecture, rather than looking for new examples of sustainable building and architectural integration.

The presented projects are examples of a study of Green Eco Solar Buildings and Seaside Village Project, developed within the ABITA Research Centre of the University of Florence, for this area, developed together with the De Feng Lida Group from Beijing, trying to harmonize traditional Chinese approach with sustainable technologies and renewable energy integration in building.

M. Sala • A. Trombadore (✉)
ABITA Interuniversity Research Centre, University of Florence, Florence, Italy

Department of Architecture - Environmental Design and Technology, University of Florence, Via San Niccolò 93, 50125 Firenze, Italy
e-mail: marco_sala@unifi.it; antonella.trombadore@unifi.it; http://web.taed.unifi.it/abitaweb/

A. Sayigh (ed.), *Seaside Building Design: Principles and Practice*,
Innovative Renewable Energy, https://doi.org/10.1007/978-3-319-67949-5_10

10.2 The Model of Green Solar Buildings

What is, however, so special about the Green Eco Solar Buildings?

The added value of the project is both the *conceptual and design approach*, as a replicable model becoming a winning model for community living and sustainable target as a profitable facility within "green living" and "green business" vision.

The completely newly defined and conceived planning for the real estate compound of Green Eco Solar Buildings will derive from a new vision oriented on sustainable development. Implementation demanded not only new procedures but also the definition of new standards and quality determinants in almost every area covered by conventional urban planning.

The energy efficiency standards applied in all the Green Eco Solar Buildings will make it possible to realize nearly zero-energy building performance. The following are the five topics for the best path for an Eco-efficient advanced seaside real estate development, fostering sustainable, cultural, and environmental local resources:

1. Ecological optimization of Green Eco Solar Building project according to the urban development and landscaper planning
2. Concept design of Green Eco Solar buildings (homes, hotels, and infrastructures)
3. Building integration of highly efficient strategies, high-tech solutions, and construction materials deriving from recycling and reuse program
4. Sustainable approach to foster local economic development
5. Integrated activities to optimize the use of renewable energy

The first step is the elaboration of urban and landscape concept for sustainable community development and the master plan of the Green Eco village in the Weihai City, with residential, hotel, commercial, and community activities and high-quality standards (Figs. 10.1 and 10.2).

Concept Design and Technologies

- *Environmental quality and renewables*
  - "Solar city": energy use optimization, district heating-cooling system, low-energy buildings, renewable energy for electricity
  - Saving measures, solar district heating system, passive solar houses, fuel cell cogeneration plant, wind turbines, photovoltaic cell technology (Fig. 10.3)
- *Construction praxis*
  - Compact layout
  - Resource-efficient layout
  - Mixed use: residential/commercial/industrial
  - Consultative and participatory planning procedures
  - Traffic minimalization concept: tram route, all amenities within easy walking distance, cycle priority route, parking space restrictions

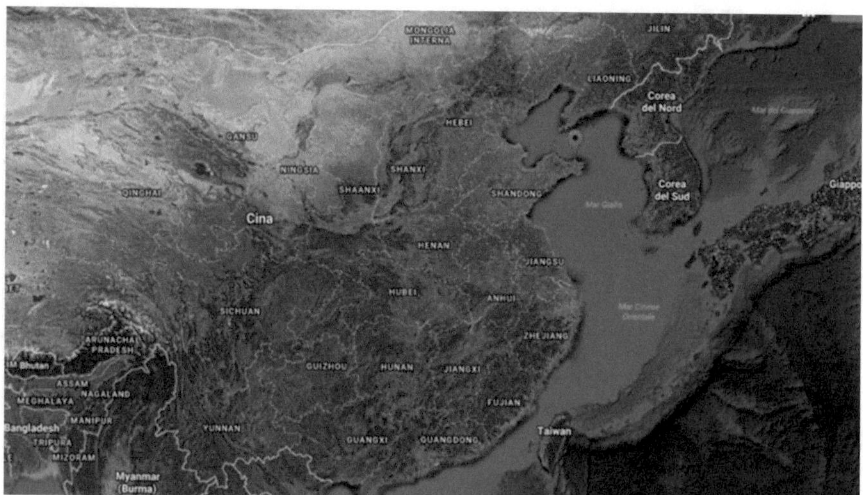

Fig. 10.1 View of the Weihai location referring to China's geographic position

Fig. 10.2 Sea site new district with residential green buildings in Weihai

  - Open-space quality: courtyards, avenues, neighborhood parks, green corridors, district park

- *Ecological standards*
  - Water: rainwater management concept as drinking water economy measures
  - Waste: ecologically compatible building materials, building waste concept, domestic and commercial waste concept

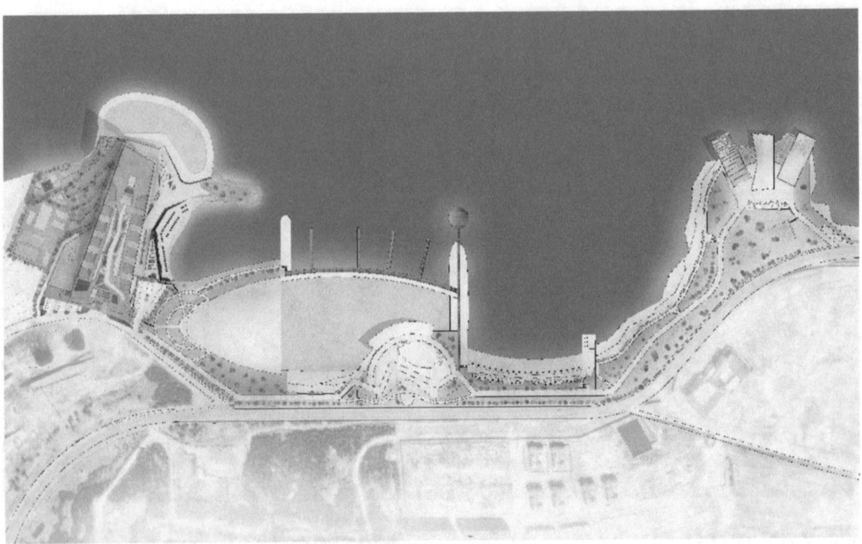

Fig. 10.3 The new master plan of sea site new district with hotel, residential, and green buildings in Weihai

- Soil: soil management, inherited pollution – removal or containment
- Landscape: ecological landscaping, ecological farming
- Environmental communications: environmental liaison agency of Green Eco Solar village
- BioCenter and Advanced Agricultural Production Center

• *Sociocultural considerations*

- Social mix of future residents
- Central tourism facilities: arts, community and advice centre, health centre
- Social infrastructure: "kinder house" with community bakery, kindergartens, decentralized support for senior citizens, space allocation for community use
- Nutrition: market for the local agricultural production distribution

According to the climatic condition, it is possible to define alternative layout of building, in order to optimize solar radiation contribution and the benefit of natural ventilation for passive cooling (Figs. 10.4 and 10.5).

In this project, to reduce the electricity needs, we integrated in the building envelope of a system (Trombadore and Scalpellini 2013):

• *The building roof*: 700 sq. of glass PV panels are integrated in the tilted top floor, and this system produced 50,00 kWp.
• *South, east, and west facade*: 150 sq. of glass PV panels are integrated in the external blade of the double skin facade (2.92 × 1.48 m), and this system produced 18,00 kWp.

Fig. 10.4 View of the central area of the new sea site district with residential green buildings in Weihai

Fig. 10.5 The green building envelope and integration of passive cooling strategies in the new sea site district in Weihai

10.3 Green Villas and Solar House Buildings

The second project is related to the master plan of the new smart green residential area and the concept design of villas and private houses. The project is based on the following strategies:

- *Energy management – smart grid and architectural integration of renewables*

The concept focuses on the integration of renewable energies: the use of specific software (as ecotect) will ensure a correct evaluation of solar contribution and the best design of PV technologies (Sala 2003). The energy management of the area is entrusted to the ICT technologies. All electrical equipment, appliances, and components are connected with a computer control system.

The control system shows the consumption of gas, water, and electricity and it also indicates any possibility of saving. The design of Eco Buildings should be constantly evolving with the aim of becoming a "world-class environment" for technology companies. The concept will develop innovative solar technologies and high-technology systems, such as solar panels, building integrated and ultra-thin, with the aim to stimulate the cooperation between companies, suppliers, scientific communities and local government.

- *The use of local construction materials and high technology*

The project will improve the use of local materials for construction buildings, as well as low-energy materials and technology, integrated with natural fiber and recycling steel and second cycle materials. Sustainable approach will be applied to foster local economic development with recycling technologies (Figs. 10.6 and 10.7).

- *Water saving and optimization*

Fig. 10.6 The view of green villas

Fig. 10.7 The green building envelope and integration of passive cooling strategies in the new sea site district

Using superficial and underwater reservoirs for rainwater treatment and bio-fito depuration together with green roofing will complete compound landscape. Water refurbishment will be done through "storm-water eco-management strategy" meaning capillary reuse purification and stoking.

In order to reduce potable water use for irrigation and domestic purposes, the project will provide all taps and possible showers with water-saving devices, as well as water-saving toilets. It is also recommended to use rainwater collected from building roofing to irrigate the reserved green areas, to wash paved areas, and to fill cisterns; the advantage in adopting these systems is increased in case sports halls and swimming pools are used.

Technological water-saving solutions are essential in this type of buildings in order to avoid potable water consumption when not necessary. Purified water can be reused to irrigate reserved green areas of hydroponic culture, ensuring the greening also when school activities are suspended (Figs. 10.8 and 10.9).

10.4 Eco Green Sea Site Hotel

The project is focalized to implement a concept design and a building construction based on technological solutions for passive cooling, natural lighting and ventilation, renewable energies, geothermal systems, greenhouses, and a smart skin envelope. The aim is to apply and test new and innovative energy-saving technologies in

Fig. 10.8 The green building with the use of local materials for building construction

Fig. 10.9 The general view of the Eco Green Sea Site Hotel

order to improve the energy performance and the indoor environment of office buildings, in order to realize a 50% reduction energy consumption and CO_2 emission and to set a new standard for energy consumption and CO_2 emissions in offices, while fully maintaining comfort conditions for workers.

Besides, the project will demonstrate that energy-efficient and sustainable buildings can fully meet all the architectural, functional, comfort, control, and safety features required, through the application of innovative and intelligent and integrated design (ARUP 2014). This demonstration could contribute to greater acceptance of innovative and renewable technologies in public buildings, improving the

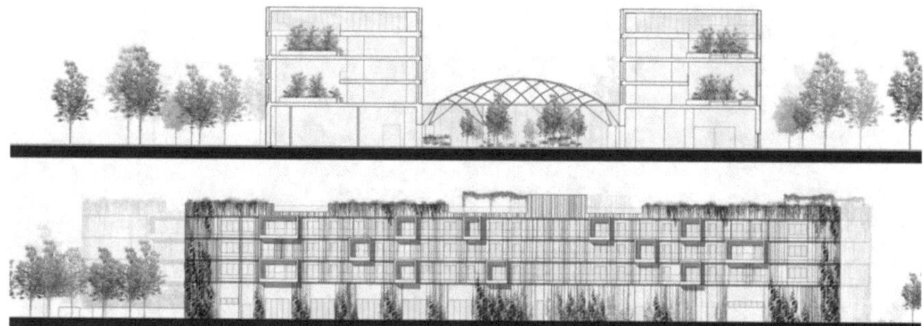

Figs. 10.10 and 10.11 The double green envelope, the central green space, and integration of passive cooling strategies

diffusion and the capitalization of the results to increase the awareness of energy-saving practices (in the medium and long term) and to integrate and improve policies for energy performance in buildings in China.

The application of a monitoring system will allow to compare the energy performance of buildings and to define a new energy standard, in accordance with EU and International Energy Labels in the Greenhouse (Figs. 10.10 and 10.11).

In particular our design the double skin façade is characterized by the external skin made with mobile vertical wood blind. This type of blinds can rotate on its vertical axis so to adjust the indoor temperature. An intermediate layer will contain the hydroponic cultures, with the plants located near the external skin so to improve their sunshine performance.

The inside envelope will be made with an inner skin done with dry assembling multilayer panels in line with external corridors with wide window development. The active facade will allow to ensure optimal energy performance in winter months with values of U of 0.8 W/m^2 K. During summer months, solar control device will be open rotating on its axis allowing the internal ventilation in the ventilated facade e microclimate regulation also in function of the hydro-ponic ventilation (Figs. 10.12 and 10.13).

The construction of this new experimental Green Hotel building incorporates energy-efficient measures, which are not only current innovations for the actual state of art but are also replicable in other building in China. The importance of this project is the introduction of some specific, innovative, energy-saving techniques, which set new energy, environmental, and health standards for industrial buildings (Trombadore 2015). In this framework, the following measures have been incorporated:

- An appropriate orientation of the building and envelope design
- Architectural integration of renewable energy, in the form of photovoltaic and solar thermal panels in the building envelope
- Smart facades – double-skin facade and highly insulated building
- Hydroponic culture inside the buffer zone of double-skin façade and in the roof of the building, so to have the farming in the building

Fig. 10.12 The plan of the new sea site Green Hotel

- The use of natural ventilation thought a good design
- The use of shading device for even distribution of daylight inside the office
- The use of a geothermal system with a heat pump for cooling and heating

10.5 Smart Dynamic Skin Building

The last project is oriented to define a concept design of a new building characterized by a smart dynamic skin. The solutions are based on technological innovation and high-performance material, according to the climatic condition during winter and summer season, fostering passive cooling systems, natural lighting and ventilation, as well as architectural integration of renewable energies (Sala and Romano 2013).

- *Double-skin facade*

Double-skin facades can be considered a technological evolution of the "curtain wall" following its rules of mounting; these systems are a solution for architectural approaches which aim at reducing energy consumption (Figs. 10.14 and 10.15). The technological characteristics of a similar facade are an internal and an external skin, a hollow space within which air flows, and other layers, creating a multiple facade system where particularly important are the speed and the flow of the air (captured both outside and inside of the building) that are recorded in the hollow space because in this way we can exploit natural ventilation in those buildings which cannot benefit from this possibility (Sala and Romano 2015). There are some advantages in the

Fig. 10.13 The view of Green Hotel – green roof as public space

use of a similar facade in China climatic condition: an easy installation, protection of the building from atmospheric agents, elimination of heat bridges, increase of the internal heat during winter, possibility of ventilation during the night in order to reduce the temperature of the rooms and refresh the internal still structures, and maximum exploitation of natural light (Fig. 10.16).

10.6 Conclusion

Throughout the design procedure, sustainable development was adapted related to the following items:

- An appropriate orientation of the building and envelope design
- Architectural integration of renewable energy, in the form of photovoltaic and solar thermal panels in the building envelope

Figs. 10.14 and 10.15 General view of the smart dynamic skin building: the new sea site in Weihai

Fig. 10.16 General view of the commercial building

- Smart facades – double-skin facade and highly insulated building
- Hydroponic culture inside the buffer zone of double-skin facade and in the roof of the building, so to have the farming in the building
- The use of natural ventilation thought a good design
- The use of shading device for even distribution of daylight inside the office
- The use of a geothermal system with a heat pump for cooling and heating

One of the new features in the author's design, the double-skin facade is characterized by the external skin made with mobile vertical wood blind. This type of blinds can rotate on its vertical axis so to adjust the indoor temperature. An intermediate layer will contain the hydroponic cultures, with the plants located near the external skin so to improve their sunshine performance.

References

ARUP. (2014). *Cities Alive – Rethinking green infrastructures*. Foresight.

Sala, M. (a cura di). (2003). *Integrazione architettonica del fotovoltaico Casi studio di edifici pubblici in Toscana*. Firenze: Alinea editrice.

Sala, M., & Romano, R. (2013). Innovative dynamic building component for the Mediterranean area. In: International scientific conference CleanTech for sustainable buildings from nano to urban scale CISBAT 2013, Lausanne, Switzerland, 4–6 September 2013, Solar Energy and Building Physics Laboratory (LESO-PB) Ecole Polytechnique Fédérale de Lausanne (EPFL) Station 18, CH-1015 Lausanne, pp. 267–272.

Sala, M., & Romano, R. (2015). Smart envelopes for Mediterranean area. The new façade system in the ICT Centre in Lucca. In L. Aelenei, M. Brzezicki, U. Knaack, A. Luible, M. Perino, & F. Wellershoff (Eds.), *Adaptive facade network – Europe* (pp. 68–68). Delft: TU Delft.

Trombadore, A. (2015). *Mediterranean smart cities*. Firenze: Edizioni Altralinea.

Trombadore, A., & Scalpellini, L. (2013). Smart skin and architectural integration of PV glazed brick shading devices. In *CISBAT 2013 CleanTech for smart cities & buildings: From nano to urban scale, proceedings* (pp. 61–65). Lausanne: EPFL. 4–6 September 2013.

Chapter 11
Conclusion

Ali Sayigh

These ten chapters highlight the various approaches adopted by the authors taking into account the particular climate zones of the areas in which they were constructing seaside dwellings. Aspects that had to be given urgent attention were the requirement for building sustainability, comfort, and mitigation of the very real impact of climate change and the importance of the utilisation of renewable energy.

Each chapter details the unique solutions that have been adopted and provides a blueprint for other architects to follow in similar circumstances.

A. Sayigh (✉)
World Renewable Energy Congress and Network, Brighton, UK
e-mail: asayigh@wrenuk.co.uk

A. Sayigh (ed.), *Seaside Building Design: Principles and Practice*,
Innovative Renewable Energy, https://doi.org/10.1007/978-3-319-67949-5_11

About the Editor

Professor Ali Sayigh, BSc, AWP, DIC, PhD, graduated from the University of London and Imperial College in 1966. He is fellow of the Energy Institute and the Institution of Engineering and Technology, chartered engineer, and chairman of Iraq Energy Institute.

Prof Sayigh taught in Iraq, Saudi Arabia, Kuwait, and England, specifically in the University of Reading and the University of Hertfordshire, from 1966 to 2004. He was head of the Energy Department at Kuwait Institute for Scientific Research (KISR) and expert in renewable energy at OAPEC, Kuwait, from 1981 to 1985.

He started working in solar energy in September 1969. Together with some colleagues, he established as an editor-in-chief *The Journal of Engineering Sciences* in Riyadh, Saudi Arabia, in 1972 and the *International Journal for Solar* and Wind Technology in 1984. The latter became *Journal of Renewable Energy* in 1990. He is editor of several international journals published in Morocco, Iran, Bangladesh, Nigeria, and India and established the World Renewable Energy Congress/Network (WREC/WREN) in 1990. He is member of various societies related to climate change and renewable energy and chairman of Iraq Energy Institute since 2010.

He was consultant to many national and international organizations, among which are the British Council, ISESCO, UNESCO, UNDP, ESCWA, UNIDO, and UN. He organized conferences and seminars in 52 different countries; published more than 600 papers; edited, written, and was associated with more than 75 books; and supervised more than 80 MSc and 35 PhD students. He has been editor-in-chief of the yearly published *Renewable Energy Magazine* since 2000; is the founder of WREN, a Renewable Energy Journal published by Elsevier; and was editor-in-chief for 30 years from 1984 to 2014.

A. Sayigh (ed.), *Seaside Building Design: Principles and Practice*,
Innovative Renewable Energy, https://doi.org/10.1007/978-3-319-67949-5

He is editor-in-chief of Comprehensive Renewable Energy coordinating contributions of 154 top scientists, engineers, and researchers in eight volumes published in 2012 by Elsevier which won the 2013 PROSE award in the USA. He founded Med Green Buildings and Renewable Energy Forum in 2011. In 2016 he established the peer-reviewed international open-access journal *Renewable Energy and Environmental Sustainability* (REES), which is published in English online by EDP in Paris.

Index

A. Sayigh (ed.), *Seaside Building Design: Principles and Practice*, Innovative Renewable Energy, https://doi.org/10.1007/978-3-319-67949-5

H

I

J

K

L

M

N

R

S

T

Printed in the United States
By Bookmasters